T0076261

MECHANICAL AND DYNAMICAL PRINCIPLES OF PROTEIN NANOMOTORS: THE KEY TO NANO-ENGINEERING APPLICATIONS

NANOTECHNOLOGY SCIENCE AND TECHNOLOGY SERIES

Safe Nanotechnology
Arthur J. Cornwelle
2009. ISBN: 978-1-60692-662-8

National Nanotechnology Initiative: Assessment and Recommendations
Jerrod W. Kleike (Editor)
2009. ISBN: 978-1-60692-727-4

Nanotechnology Research Collection - 2009/2010. DVD edition
James N. Ling (Editor)
2009. ISBN: 978-1-60741-293-9

Nanotechnology Research Collection - 2009/2010. PDF edition
James N. Ling (Editor)
2009. ISBN: 978-1-60741-292-2

Strategic Plan for NIOSH Nanotechnology Research and Guidance
Martin W. Lang
2009. ISBN: 978-1-60692-678-9

Safe Nanotechnology in the Workplace
Nathan I. Bialor (Editor)
2009. ISBN: 978-1-60692-679-6

Nanotechnology in the USA: Developments, Policies and Issues
Carl H. Jennings (Editor)
2009. ISBN: 978-1-60692-800-4

Electrospun Nanofibers Research: Recent Developments
A.K. Haghi (Editor)
2009. ISBN: 978-1-60741-834-4

Nanofibers: Fabrication, Performance, and Applications
W. N. Chang (Editor)
2009. ISBN: 978-1-60741-947-1

Nanofibers: Fabrication, Performance, and Applications
W. N. Chang (Editor)
2009. ISBN: 978-1-61668-288-0 (Online Book)

Nanopowders and Nanocoatings: Production, Properties and Applications
V. F. Cotler (Editor)
2010. ISBN: 978-1-60741-940-2

Barrier Properties of Polymer Clay Nanocomposites
Vikas Mittal (Editor)
2010. ISBN: 978-1-60876-021-3

Bio-Inspired Nanomaterials and Nanotechnology
Yong Zhou (Editor)
2009. ISBN: 978-1-60876-105-0

Nanomaterials Yearbook - 2009. From Nanostructures, Nanomaterials and Nanotechnologies to Nanoindustry
Gennady E. Zaikov andVladimir I. Kodolov (Editors)
2010. ISBN: 978-1-60876-451-8

Nanomaterials: Properties, Preparation and Processes
Vinicius Cabral and Renan Silva (Editors)
2010. ISBN: 978-1-60876-627-7

Nanotechnology: Nanofabrication, Patterning and Self Assembly
Editors: Charles J. Dixon and Ollin W. Curtines
2010. ISBN: 978-1-60692-162-3

Mechanical and Dynamical Principles of Protein Nanomotors:
The Key to Nano-Engineering Applications
A. R. Khataee and H. R. Khataee
2010. ISBN: 978-1-60876-734-2

NANOTECHNOLOGY SCIENCE AND TECHNOLOGY SERIES

MECHANICAL AND DYNAMICAL PRINCIPLES OF PROTEIN NANOMOTORS: THE KEY TO NANO-ENGINEERING APPLICATIONS

ALI R. KHATAEE

AND

HAMID R. KHATAEE

Nova Science Publishers, Inc.
New York

LIBRARY OF CONGRESS CATALOGING-IN-PUBLICATION DATA

Khataee, A. R. (Ali Reza), 1977-
Mechanical and dynamical principles of protein nanomotors : the key to nano-engineering applications / authors, A.R. Khataee and H.R. Khataee.
    p. cm.
Includes bibliographical references and index.
ISBN 978-1-60876-734-2 (softcover : alk. paper)
    1. Protein engineering. 2. Proteins--Mechanical properties. 3. Proteins--Industrial applications. 4. Nanostructured materials. 5. Motors. I. Khataee, H. R. II. Title.
TP248.65.P76K485 2009
660.6'3--dc22
                2009044349

Published by Nova Science Publishers, Inc. ✦ *New York*

CONTENTS

PREFACE

It is obvious that movement is an essential concept of all living organisms. Molecular motility participates in many cellular functions including cell division, intracellular transport and movement of the organism itself. Thus, it is not surprising that nature has evolved a series of biological nanomotors that fulfill many of these tasks. A general class of these biological nanomotors is called protein nanomotors that move in a linear fashion (e.g. the kinesin or myosin or dynein motors) or rotate (e.g. F_0F_1-ATP synthase or bacterial flagellar motors). Protein nanomotors are natural motors responsible for the human activity and are also the subject of interest for nanotechnology. Protein nanomotors are ideal nanomotors because of their small size, perfect structure, smart and high efficiency. Recent advances in understanding how protein nanomotors work has raised the possibility that they might find applications as protein-based nanorobots. Thus bio-nanomotors could form the basis of bottom-up approaches for constructing active structuring and maintenance at the nanometer scale. In this chapter, we have presented structures, mechanisms and potential applications of linear protein nanomotors. The three known families of protein nanomotors kinesin, dynein and myosin are multi-protein complexes and share a variety of important features. They are responsible for various dynamical processes for transporting single molecules over small distances to cell movement and growth. Our reviewing from the mechanism, regulation and coordination of linear nanomotors, indicate that the majority of active transport in the cell is driven by linear protein nanomotors. All of them convert the chemical energy into mechanical work directly rather than via an intermediate energy. Linear protein nanomotors are self-guiding systems. They have evolved to enable movement on their polymer filaments, either on cellular or supra-cellular levels and are able to recognize the direction of movement. Moreover, each class of nanomotor has different properties, but in the cell they are known to cooperate and even to compete with each others during their function. We have also reviewed the potential application of linear protein nanomotors. According to this, we predict that linear protein nanomotors may enable the creation of a new class of nanotechnology-based applications; for example, bio-nanorobots, molecular machines, nanomechanical devices and drug deliver systems. Thus, protein nanomotors field is very challenging field and is attracting a diverse group of researchers keen to find more.

ACKNOWLEDGEMENTS

We are grateful to the University of Tabriz, Iran for financial and related supports. We kindly thank Dr. M. H. Rasoulifard (Department of Chemistry, Faculty of Science, University of Zanjan, Zanjan, Iran) and Dr. M. Zarei (Department of Applied Chemistry, Faculty of Chemistry, University of Tabriz, Tabriz, Iran) for valuable scientific advice.

INTRODUCTION

It is obvious that movement, in one form or another, is an essential feature of all life at both the macroscopic and cellular level. Organisms, from human beings to bacteria, move to adapt to changes in their environments, navigating toward food and away from danger. By evolutionary modification over billion of generations, living organisms have perfected an armory of biological nanomachines, structures, and processes. Cells, themselves, are not static but are bustling assemblies of moving proteins, nucleic acids, and organelles. Therefore, life is made possible by the action of a series of biological nanomachines in the cell machinery. Realisation of biological nanomachines would pave the way for novel devices and processes capable of revolutionising medicine, or of reducing resource consumption and environmental pollution in manufacturing processes. This would bring enormous benefits in terms of human health and quality of life, as well as greatly enhancing the industrial competitiveness of any region able to gain a lead in the underlying technologies.

The concept of molecular engineering was first introduced by Richard Feynman in 1960 [1]. Norio Taniguchi was the first to define nanotechnology in 1974 [2]. The combination of cellular-molecular biology and nanotechnology has led to a new generation of nanoscale-based devices and methods for probing the cell machinery and elucidating intimate life processes occurring at the molecular level that were heretofore invisible to human inquiry. Therefore, automatic movement of the nano particles (or nanomotor) is still under development and the field is under maturation in this decade [3].

Generally speaking, a motor may be defined as a device that consume energy in one form and converts it into motion or mechanical work; for example, any biological nanomotor. Most of biological nanomotors are based on either eukaryotic enzymes or mechanoenzymes. The eukaryotic enzymes move in approximately straight lines along cytoskeletal filaments while the mechanoenzymes employs a rotary mechanism, in either the clockwise (cw) or counterclockwise (ccw) direction. Eukaryotic enzymes take the energy from hydrolysis of adenosine-5′-triphosphate (ATP), Nature's universal energy currency, and mechanoenzymes take their energy directly from the proton gradient across the cytoplasmic membrane. Eukaryotic cells contain three major families of protein nanomotors: myosins, kinesins, and dyneins [4]. These three linear protein nanomotors are powered by the hydrolysis of ATP through which they convert chemical energy into mechanical work [5]. ATP is a small nucleotide molecule that acts as a common bioenergy fuel in the cell. The carbon–nitrogen bond in ATP is highly energetic, and during the hydrolysis, $ATP \rightarrow ADP + P_i$, a phosphate group is decoupled from the ATP molecule resulting in the appearance of the reaction products, adenosine-5′-diphosphate (ADP) and inorganic phosphate (P_i), together with the release of about 20 $k_B T$ energy [6]. Most of linear protein nanomotors are able to pull vesicles, organelles and other types of cargo over large distances, from micrometers up to meters, along the surface of a suitable substrate. These nanomotors are able to recognize the rail polarity and so the direction of transport [7]. Differences in function are observed between these classes of nanomotors, but in the cell they often function together [8].

Biology provides brilliantly developed set of examples: in living systems, nanomotors do exist, and they do perform extraordinarily sophisticated functions. Biological nanomotors are rather new and are attracting a diverse group of researchers keen to find more. We have recently reviewed the beautiful highly sophisticated F_0F_1-ATP synthase biological protein nanomotor [9]. In this chapter we explain recent progresses in linear protein nanomotors including kinesins, dyneins and myosins. The organization of this chapter is as follow: In section 2, we have described structures of these nanomotors and also pointed to their respective filaments, briefly. In section 3, we have explained the function of nanomotors with other aspects of their function. Finally, in section 4, we have provided examples of emerging applications of these nanomotors. Henceforth, unless otherwise mentioned, 'protein nanomotors' will refer to the linear protein nanomotors.

STRUCTURE OF
THE LINEAR PROTEIN NANOMOTORS

At first glance, three major families of protein nanomotors appear to be quite different from one another. Many protein nanomotors are dimers with two 'heads' connected together at a 'stalk' region and a 'tail' domain opposite the heads. The head of the nanomotors contains the motor domain that provides the motion along the filaments whereas the tail of the nanomotors contains the subunits responsible for cargo binding and regulation [10]. Orderly motion of these nanomotors across distances requires a filament that steers the motion of the nanomotor assembly. Therefore, before describing structures of the three nanomotors, we point to the respective filaments including microtubules (MTs) and actin filaments. MTs and actin filaments form a network of highways within cells, and localized cues are used to target specific cargos to specific sites in the cell [11].

MTs have a crucial organizing role in all eukaryotic cells (see Figure 1). MTs are nucleated and organized by the microtubule organizing centers (MTOCs), such as centrosomes and basal bodies, situated close to the nucleus. A microtubule is a hollow cylindrical array of $\alpha\beta$-tubulin dimers arranged in a head-to-tail configuration [12]. The outer diameter of MTs is 25 nm [13, 14]. MTs have highly dynamic instability [14, 15]. A MT is a polar structure, whose ends are distinct in both structure and behavior [12, 14]. One end, termed the minus-end, is anchored near the center of a cell, whereas the other, so called plus-end extends toward the cell surface and this leads to a radial organization of the MT network in some interphase cells, such as fibroblast cells [16], pigment cells [17, 18] and certain mammalian cells [19]. MT organization is cell-type specific and in some cases, such as neurons [20] and epithelial cells [21], differs significantly from the mentioned radial organization.

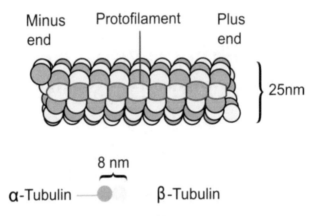

Figure 1. Schematic of the MT: MTs are built from two kinds of homologous subunits, α- and β-tubulin, which assemble in an helical array of alternating tubulin types to form the wall of a hollow cylinder. They are highly dynamic structures that grow through the addition of α- and β-tubulin to the ends of existing structures.

Actin filaments (thin filaments) are found in all eucaryotic cells and are essential for many movements within them, especially those involving the cell surface (Figure 2). Actin monomers (G-actins) come together to form actin filaments (F-actins) [22]. Like a MT an actin filament has a structural polarity, with a plus-end and a minus-end [23]. Like many biological structures, actin filaments are self-assembled. Actin filaments are thinner, more flexible, and usually shorter than MTs [17, 24]. They are generally found in cross-linked bundles and have been suggested to bridge the cap between MTs, for example in cultured rat axons [25, 26]. In this way, local transport can occur on actin filaments in regions with few MTs, as at the axon terminal [26]. In some cases, actin filaments have an ordered structure close to the cells surface with barbed (plus) ends pointed outwards, thus allowing their nanomotors to transport cargos to the very edge of the cell [27, 28]. At least in some cases, however, further inside cells the actin filament network is approximately randomly oriented and has sufficient density to make it a good local transport system [27, 29]. This random distribution of actin filaments can be used to spread out cargos [29], enabling the cell to achieve a more uniform distribution of cargos than would be possible by moving on MTs alone [18, 30].

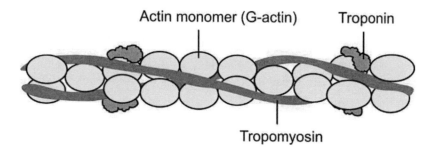

Figure 2. Schematic of the actin filament: Actin is a part of the contractile apparatus of the muscle and a part of the cytoskeleton of individual cells. These thin actin filaments are closely associated with tropomyosin and troponin. Tropomyosin is a filament that lies alongside of actin in the thin filament. Spaced at regular intervals along the tropomyosin filament are troponin complexes.

2.1. STRUCTURE OF THE KINESIN PROTEIN NANOMOTOR

Kinesin is a protein nanomotor associated with MTs identified in 1985 as the force that underlies the movement of particles along MTs within the giant axon in squid [31-33]. Conventional first discovered kinesin is a protein dimer consisting of two heavy chains and two light chains [34]. Each of its heavy chains contains a globular head connected via a short, flexible neck linker to the long, central coiled-coil stalk region that ends in a tail region. The cargo binding region resides on two light chains associated with the heavy chains (see Figure 3) [35, 36]. In most of kinesin nanomotors, the kinesin light chains bind to the C-terminal ends of the stalk domains and heavy chains bind to the N-terminal ends of the motor domain [37]. Together, the head and neck domain forms a functional motor unit often referred to as the motor domain. In contrast to the other class of MT-dependent nanomotors, the dyneins, or the actin-dependent nanomotor myosin, kinesin's motor domain is much smaller and fully functional without additional polypeptide chains. Kinesin's globular motor domain exhibits structural similarity, but little sequence homology, to that of myosin [38]. The kinesin head contains nucleotide binding site [35] and MT binding sites [39]. The MT binding site at kinesin head lies on the surface exactly opposite to the ATP binding pocket [39]. Therefore, both the ATP and MT could have simultaneous access to the head. The catalytic core of kinesin motor domain is connected to the 'neck helix' through a short flexible neck linker domain and this neck helix is linked to the coiled-coil stalk [40-42]. This neck linker changes conformation in response to nucleotide binding, instead of a long α-helical lever arm in myosin structure that will be described in section 2.3. Stalk domain of the heavy chain interacts in an α-helical coiled-coil that extends from the heavy chain neck to the tail. The coiled-coil stalk is interrupted by a few hinge regions that give flexibility to the otherwise stiff stalk domain. The neck and stalk domains are linked by the unstructured hinge 1 domain [42]. The stalk is comprised of two coiled-coil domains that are linked by the hinge 2 domain [43-45]. Kinesin tail domain interacts with the cargos and transports them on MT to a variety of destinations inside the cell.

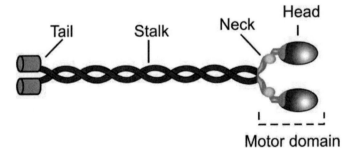

Figure 3. A simplified structure of kinesin: typical kinesin posses' a three-domain organization: a motor domain, a coiled-coil stalk domain and a globular tail domain. The stalks intertwine to form the kinesin dimmer from the heavy chain neck to the tail. Cargo binds to the tail while the twin heads alternately bind the MT as the kinesin pulls the cargo along.

2.2. STRUCTURE OF THE DYNEIN PROTEIN NANOMOTOR

Dynein has a rather different structure in comparison to kinesin and myosin. In other words, as can be seen from Figure 4, dynein appears to be a much more complex nanomotor by construction [46, 47]. Each dynein molecule consists of a dimmer of heavy chains (HCs) that contain the motor domain as well as a variety of accessory proteins termed intermediate chains (ICs), light intermediate chains (LICs) and light chains (LCs) [46, 48, 49]. In contrast to both kinesin and myosin, the heavy chain of dynein is a very large molecule with a molecule weight greater than 500 kDa. Sequence analysis studies of dynein show that it belongs to the AAA (ATPase associated with various cellular activities) class of proteins, which makes the structure of dynein fundamentally different from that of kinesin or myosin [47, 50]. In contrast to the single ATP binding in kinesin and myosin heads, dynein has multiple ATP binding sites in each head [51-53]. Electron microscopy studies have shown that the dynein head has seven globular domains, out of which six are AAA domains, arranged in a ring-like conformation around a central cavity [54]. AAA5 and AAA6 are not capable to bind ATP, whereas AAA1– AAA4 can bind ATP, though with varying binding affinities [51-53, 55-57]. AAA1 is the primary site of ATP hydrolysis, although all four functional sites seem to cooperate [57, 58]. A possible biological importance of such complexity and multiple ATP binding sites is a proposed gear mechanism within the dynein head [59]. The heavy chain folds into this ring-like structure composed of multiple ATP binding domains as well as a stem and a stalk that emerges from the head domain [49]. The coiled-coil stalk, binds to and walks along the surface of the MT via a repeated cycle of detachment and reattachment sequence. However, the extended tail (also called stem), binds to the intermediate and light chain subunits which attach the dynein to its cargo [60]. The conventional view of dynein nanomotors, in general, is that the stem emerges from the globular head at a position opposite to the MT-stalk. This architecture is distinctly different from the single globular domains of the motility-related kinesin and myosin proteins [61-63]. The stem connects to the head by means of a linker approximately 10 nm long that lies across the head [60]. The linker domain has been suggested to interact with the AAA ring and has been proposed to be involved in force generation [60-64]. Various appearances have long suggested that the stem is highly flexible [65, 66]. The stalk takes the flexible structure of a 13 nm long coiled-coil [67]. The coild-coil stalk emerges after AAA4 and contains the MT binding domain that is located at the tip of the coiled-coil stalk [67-69].

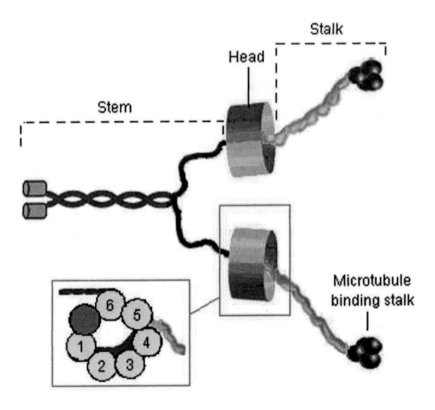

Figure 4. A simplified structure of dynein: the dynein heavy chain is the dynein nanomotor subunit, composed of three domains: stem, head, and stalk. A single dynein head is composed of a ring-shaped arrangement of six ATP binding domains, with a protruding stalk that engages MTs (MT-stalk) and a tail (or stem) that connects to cargo. The cylindrical shapes on the left symbolize the cargo binding domain.

2.3. STRUCTURE OF THE MYOSIN PROTEIN NANOMOTOR

The results of electron microscopy studies of skeletal muscle myosin reveals that it is a two-headed structure linked to a long stalk (Figure 5). Similar to kinesin, myosin consists of a homodimer of heavy chains and a tail consisting of coiled-coil light chains. Each heavy chain forms myosin head which is the actin binding motor domain, and a long alpha helix tail. Most of myosin nanomotors, similar to kinesin nanomotors, share N-terminal head domains that mediate the ATP-sensitive binding to actin filaments and C-terminal tail domains [37]. Several functionally important landmarks exist on the myosin structure. Near the midpoint of the long linear superhelical region is a site defined by its ready susceptibility to proteolytic trypsin digestion. Trypsin cleaves myosin into 2 portions: the portion containing the headpiece is known as heavy meromyosin (HMM) and other portion is known as light meromyosin (LMM). A second proteolytic landmark susceptible to papain has also been considered a hinge point (See Figure 5). Papain cleaves the HMM into two subfragments 1 (S1s) and a subfragment 2 (S2) [70]. Each S1 fragment includes the head from the heavy chain and one copy of each light chain. Each of trypsin and papain acts as a hinge point in myosin structure. The myosin head has several important characteristics: (1) it has ATP binding sites into which ATP molecules fit, (2) it has actin binding sites into which fit molecules of actin, equivalent to the position and interaction of the MT binding site of kinesin, and (3) it has a hinge at the point where it leaves the core of the myosin polymer (thick filament), that will be defined in section 3.3, and this allows the head to swivel back and forth [71]. All of the myosin motor domains with known structures consist of a small compact subdomain converter. Also the lever arm is a part of the long alpha helix. Crystal structures of myosin II, a muscle myosin [72], have shown [73-76] that small movements within the myosin motor core are transmitted through the converter domain to the lever arm consisting of a light-chain-binding helix and associated light chains [75, 76]. This indicates that, the lever arm further amplifies the motions of the converter domain into large directed movements [73, 75-77]. Both myosin and kinesin are thought to produce force by effecting the rotation of a relatively rigid lever arm that is directly connected to the motor; the nanomotor heads pivot about this junction [78, 79].

On the basis of the above-mentioned structures of kinesin, dynein and myosin nanomotors, it can be concluded that members of the myosin and kinesin families were found to have remarkable similarities.

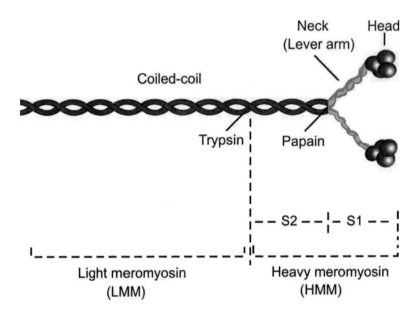

Figure 5. A simplified structure of muscle myosin: the globular domains at the right are motor domains (heads or cross bridges) that also interact with the actin filaments. Several functionally important landmarks exist on the myosin molecule. Treatment of myosin with trypsin and papain results in the formation of four fragments: two S1 fragments, an S2 fragment, also called heavy meromyosin (HMM), and a fragment called light meromyosin (LMM). The S2 and LMM fragments are largely α helical, forming two-stranded coiled-coils created by the remaining lengths of the two heavy chains wrapping around each other.

FUNCTION OF THE LINEAR
PROTEIN NANOMOTORS

Protein nanomotors are the active workhorses of the cells [80, 81]. The majority of active transport in the cell is driven by the three classes of protein nanomotors namely kinesin, dynein and myosin. The reason that biological nanomotors are necessary in cells is that diffusion is too slow to transport molecules efficiently from where they are made (typically near the nucleus) to where they are used (is often at the periphery of the cell). These nanomotors are unusual nanomachines that do what no man-made machines do: they convert chemical energy to mechanical energy directly rather than via an intermediate such as heat or electrical energy. It is obvious that the confinement of heat, for example, on the nanometer scale is not possible because of its high diffusivity in aqueous media [11]. The protein nanomotors utilizing the cytoskeleton for movement fall into two categories based on their substrates: Actin-based nanomotors such as myosin move along microfilaments through interaction with actin [71] and MT-based nanomotors such as dynein [82] and kinesin [32] move along MTs through interaction with tubulin. MT-based nanomotors are divided into two groups: plus-end nanomotors (kinesins) [83] and minus-end nanomotors (dyneins) [82], depending on the direction in which they walk along the MT cables within the cell. All family members are able to achieve movement by binding and hydrolyzing ATP in their globular motor domain. Also most of nanomotors specifically bind to a particular filament and are actively moving cargo only in one direction of the polarized filaments. Some cargos moved by such protein nanomotors are vesicles, mRNA, mitochondria, endosomes, virus particles, etc [8] (Figure 6). Thus, protein nanomotors with different architecture and function have evolved [7, 48, 84].

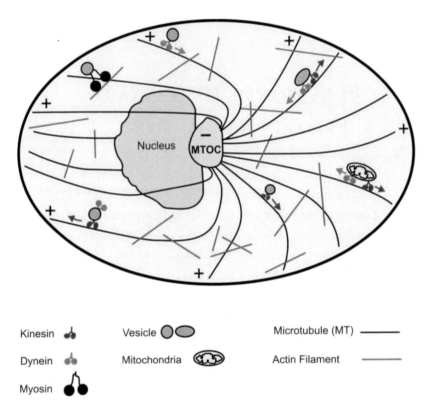

Figure 6. Schematic of a typical interphase cell. Internal order in eukaryotic cells is created by using protein nanomotors that shuttle various cargos along cytoskeletal filaments. A few different forms of cargo are being transported by MT-based nanomotor (kinesin and dynein) or an actin-based nanomotor (myosin). Multiple nanomotors, even of different kinds can attach to and transport a given cargo, usually in a bidirectional manner along the MT.

Intracellular transport occurs along filaments (MTs or actins) when the appropriate nanomotor binds to a cargo through its tail and simultaneously binds to the rail through one of its heads [85]. The nanomotor then moves along the rail by using repeated cycles of coordinated binding and unbinding of its two heads, powered by energy derived from hydrolysis of ATP [48, 61, 84, 86]. To fully understand the function of a protein nanomotor, it is important to understand the exact cycle of events and how these are coupled to the mechanical events that generate processive motion. Processivity describes the ability of a protein nanomotor to move across its polymer substrate without dissociation; a highly processive motor will execute many repetitions of its motility cycle while remaining bound to its track, while a nonprocessive motor dissociates after each step. The processivity is depicted by 'duty ratio', which indicates the relative time a motor domain remains bound on the filament, as fraction of the entire ATP hydrolysis cycle. The processive nanomotors have a duty **ratio** ≥ 0.5 [87]. Processive nanomotors are specially suited to function as vesicle transporters in the cytoplasm, because if a nanomotor detaches from the filament, the cargo is likely to diffuse away. Each of the three classes of protein nanomotors i.e. kinesins, dyneins and myosins, has members with processive motion in their families. In this section we have summarized the function of the above-mentioned protein nanomotors. Furthermore, we have compared the processivity of them and described cargo transport by several nanomotors and also pointed to the regulation and coopration of these three nanomotors.

3.1. FUNCTION OF THE KINESIN PROTEIN NANOMOTOR

Most kinesins transport cargo within the cell by walking along MT tracks. Before describing how kinesin nanomotors move along MTs, it should be noted that kinesin nanomotors exist in two differently form: inactive (folded) and active (unfolded) forms [37]. To prevent futile ATP consumption by kinesin while not carrying cargo, kinesin's activity is switched off when cargo is not bound [43]. *In vitro* ATPase and motility assays [44, 45, 88-90] and *in vivo* mutational analyses [45, 91] demonstrated that kinesin's C-terminal tail domain contains a kink at hinge 2 (folding site) that, when cargo is not bound, enables kinesin to assume a folded conformation in which the tail binds intramolecularly to the neck domain. Analysis of kinesin truncation and point mutations indicate that in this folded conformation, a motif on the tail domain inhibits ATPase activity in the head domain and diminished binding to MTs [45, 89, 92]. This may prevent wasteful hydrolysis of ATP by kinesin when it is not transporting cargo. Thus, the nanomotor remains in the inactive folded form until the binding of an appropriate cargo protein triggers a conformation change and activates ATP-dependent translocation and leads to active extended form (unfolding form) of kinesin for movement along MT.

It was generally noted that kinesins with the motor domain at the N-terminals are plus-end directed; the ones with C-terminal motor domains are minus-end directed; and the ones with motor domain in the middle are not so motile but used for breaking the MT [87]. The movement of conventional kinesin is unidirectional toward the plus-end of MTs and is known as anterograde transport [93]. The velocity of this movement is up to 3μm/s [94]. Most kinesins walk unidirectionally towards the plus-end of a MT which, in most cells, entails transporting cargo from the MTOC towards the cell periphery. However, a different kinesin-like protein was identified from *Drosophila*, called *ncd* (non-claret disjunctional), move toward the minus-end because it's C-terminal motor domain [95, 96]. For the kinesin movement many types of models have been presented [86, 97-112]. The prevailing ones propose that the dimer moves hand-over-hand and it maintains continuous attachment to MT by alternately repeating single-headed and double-headed bindings [86, 97-102, 104, 105, 107-109, 111, 112]. As can be observed from Figure 7, we briefly and generally point to the hand-over-hand model of kinesin movement. We begin with one head of a two-headed kinesin molecule, which both heads in the ADP form, binds to a MT. The binding of ATP triggers a conformational change (power stroke) in the head domain that leads to locking this head domain in place to MT and throwing the second head toward the plus end of MT. In other words, the walk generation would require swinging one head with respect to the other and alternating ATP hydrolysis at the two heads, a process which is attributed to the hinge neck region of kinesins [100]. These conformational changes move the motor domains in a hand-over-hand movement along the tubulin rail. This action of movement is very similar also for the big brother of kinesin, myosin-V, which will be described in section 3.3. Thus, to transport cargos for a sufficiently long distance, kinesins would need to do the following: (1) associate with the cargo at the beginning of the run, (2) maintain a sufficiently long run length while associated with the cargo, and (3) release the cargo at an appropriate site at the end of the run or transfer it to another nanomotor complex. How these three functions are coordinated is yet to be more understood. In a typical force-clamp experiment [113] in which a range of external forces was applied to the kinesin, it has been found that kinesin hydrolyses one ATP molecule per single step of its movement. This forms one of the main properties of the kinesin nanomotors. A further experiment using fluorescence imagining with 1 nm accuracy was performed to determine the average step size of the kinesin, and it was found that the step-size is about 8 nm [114]. Since each $\alpha\beta$-tubulin dimer of MT can bind at most one head, it appears that a single head has to cover the length of two dimers (about 16 nm) during each step. This implies that each of the heads is either stationary or move by about 16nm, while the cargo, on the centroid of the kinesin, is moved by about 8 nm. Also a single kinesin head appeared to apply an average force of 5 pN/motor domain per step [115].

It has turned out that conventional kinesins are able to move large distances along the filament as single molecules without detaching. In other words, conventional kinesin is a processive nanomotor [116-118]. *In vitro* motility assay also showed that the dimeric kinesins are highly processive nanomotors [97, 119]. Processive nanomotors, such as kinesin-1 [113, 120], kinesin-2 [121] and myosin-V [122], have a high duty ratio (about 1) so rarely detach during motion. Coordination of movements between the two motor subunits in a dimer is essential to maintain such processivity [123]. Therefore, a processive dimeric kinesin would always have one of its two heads attached to the MT during the walk. Studies with both the unconventional myosins (Myosin-V, etc.) and kinesin heavy chain suggest that the length of the neck linker region is necessary for maintaining processivity as well as the run length [124]. Like conventional kinesin, *ncd* is a dimer in solution but, in contrast, it is a nonprocessive nanomotor [125, 126]. The kinesin homodimer also exhibits an alternation in the mean dwell time, the length of time that each of the kinesin heads spends attached to the MT, in successive steps. In other words, the dwell time spent by a kinesin head, before it makes the next move, can be different for each step, implying that the kinesin performs a form of 'limping' motion during its processive movement. This indicated that the kinesin follows an asymmetric hand-over-hand type of motion [127]. Two models have been proposed to explain the limping behaviors: (1) the limping is caused by an over- and under-winding of the coiled-coil during the hand-over-hand motion, and (2) the limping is induced by an axial misregistration between the α helices of the coiled-coil dimerization domain [127].

Since its discovery, kinesin has been studied extensively by using various experimental methods [86, 98-101, 128, 129]. In particular, by using optical trapping nanometry, many aspects of its dynamics such as the mean movement velocity, randomness, mean run length, backward stepping and limping behaviors under various loads and ATP concentrations have been elaborately studied [113, 119, 127, 130-141]. The processive movement of the kinesin nanomotors is numerically simulated with a lattice model based on Monte Carlo simulation [142]. The Monte Carlo simulation results agree well with the experimental data on the relation of velocity versus ATP and ADP concentrations.

3.2. FUNCTION OF THE DYNEIN PROTEIN NANOMOTOR

Dynein powers the motion of cilia and flagella in some eukaroytic cells. Dynein also transports various cellular cargos from the periphery of the cell toward the MTOC by walking toward the minus-end of MTs, that is known retrograde transport, and its function is necessary for a wide variety of processes [46, 48, 84]. This leads to divide dynein into two groups: cytoplasmic dyneins and axonemal dyneins [85]. Cytoplasmic dyneins perform various intracellular cargo transport functions and axonemal dyneins are anchored in large linear arrays along MTs inside cilia and flagella [143].

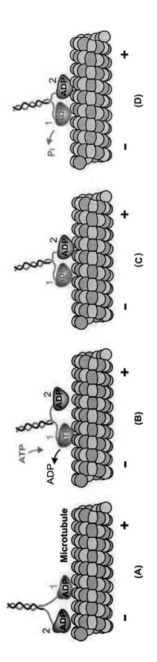

Figure 7. Schematic of kinesin movement. (A) One head of a two-headed kinesin molecule, initially with both heads in the ADP form, binds to a MT; (B) Release of ADP and binding of ATP results in a conformational change that locks the first head to the MT and pulls the neck linker to the head domain, throwing the second head toward the plus end of the MT. The resultant pull is transferred to the neck helix, which is linked to the coiled-coil stalk; (C) ATP hydrolysis occurs while the second head interacts with the MT; (D) The release of P_i and repeating the cycle.

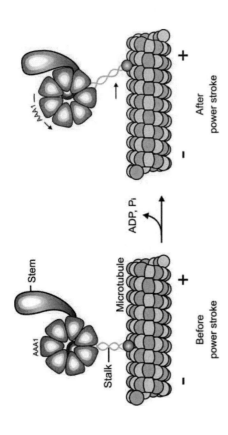

Figure 8. Schematic of the movement of dynein protein nanomotor. A superposition of two conformational states of dynein was seen for the first time before and after its power stroke. These conformations were produced by the presence and absence respectively of ADP-vanadate. In the absence of nucleotide, the stem and stalk were emerged much closer together owning to a change in linker orientation and the coiled-coil stalk was become stiffer.

Electron microscopy reconstructions of cytoplasmic dynein show a structure similar to axonemal dynein, so it is likely that the power stroke of their functions occurs in a similar way [54]. A swing of the stem and stalk is induced by ATP hydrolysis was proposed to be responsible for the power stroke of dynein nanomotor that allows movement along the MTs [60]. This power stroke is thought to accompany loss of ADP and P_i from the enzyme. To visualize the dynein power stroke, Images of axonemal dynein after locking the nanomotor into its presumed pre-power and post-power stroke conformations (respectively, in the ADP-vanadate bound state and in the absence of nucleotide) has been taken (Figure 8) [60, 144]. It was found that product release after ATP hydrolysis leads to rotation of the ring-like dynein head around the motor–stem junction. This translocates the MT by about 15 nm, proposed to be the mean value of the power stroke [144]. As can be observed in Figure 8, there are many changes in the structure when products of ATP hydrolysis are released. The most pronounced is a movement of the stem relative to the head, to bring it closer to the stalk, and this is probably caused by movement of the linker across the surface of the head. The coiled-coil stalk also stiffens, which suggests a means of transmitting information between the tip of the stalk and the ATPase site in the head. Therefore, force generation is proposed to involve a rotation of the entire head. The AAA2–AAA4 region of the ring is hypothesized to play a regulatory role in the transmission of force during dynein power stroke, and is therefore called the 'regulatory domain' of the dynein head [51, 52, 55, 57, 59, 60, 144, 145]. Several studies showed that compared with myosin and kinesin, the precise molecular details of the dynein conformational changes that induce the motion are not well understood [146, 147]. Therefore, we have firstly introduced the cytoplasmic and axonemal dynein and then focus on the limping and the processivity behaviors of dynein nanomotor.

Cytoplasmic dynein performs several functions necessary for cell survival such as organelle transport [148]. Only two members of the family are cytoplasmic, one of which, cytoplasmic dynein 1, formerly called MAP1C (microtubule associated protein 1C), is responsible for a wide range of cellular functions [149]. *In vitro* studies have shown that cytoplasmic dynein 1 is a processive nanomotor [59, 150, 151]. The second member of the cytosolic dyneins, dynein 2, is much more restricted in its functions. In cell cultures of mammalian cells it is seen associated with cilia and in the region of the Golgi apparatus [152, 153]. In tissues it seems to be limited to ciliated cells and to ciliary structures [153].

Axonemal dynein is the nanomotor that causes the ciliary and flagellar motion. Cilia and flagella share a common design (Figure 9). MTs are also the fundamental building blocks of cilia and flagella. Cilia and flagella are bounded by the plasma membrane and have a core axoneme, a complex of MTs and associated proteins [154]. The axoneme is a complex bundle of MT fibers that includes two central, separated MTs surrounded by nine pairs of joined MTs. Axonemal dynein causes sliding of MTs in the axonemes of cilia and flagella. In flagellar and ciliary axonemes, outer and inner dynein arms are projected from the A- tubule of peripheral doublet MTs. These bind to the B-tubule of neighboring doublet MTs in an ATP dependent manner. In this case, the A- tubule corresponds to their cargo. Axonemal dynein barely detaches from the cargo, A- tubule, while the cargo of cytoplasmic dynein is thought to be detached when it arrives at the cell center to recruit another nanomotor [155]. During the power stroke, which causes movement, the AAA ATPase motor domain undergoes a conformational change that causes the MT-binding stalk to pivot relative to the cargo-binding tail with the result that one MT slides relative to the other [148]. These processes cause the motion of the cilium or flagellum. Axonemal dyneins are not required to be processive since they function as a large linear array of nanomotors. However, cytoplasmic dynein probably moves processively along the MT; that is, one or the other of its stalks is always attached to the MT so that the nanomotor can walk a considerable distance along a tubule without detaching [47, 49, 63]. Thus, the targeting mechanism is different between the two dynein families.

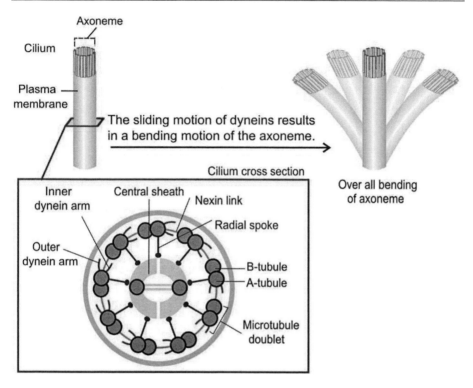

Figure 9. Mechanism of ciliary motion: Cross section of cilium or flagellum shows that the axoneme of cilia or flagella includes nine doublet MTs around the periphery; two singlet central MTs, surrounded by a sheath; nexin links and radial spokes. A basal body, which is a single centriole cylinder, is at the base of each cilium or flagellum. The smaller-diameter tubule of each doublet, which is a true cylinder, is called the A-tubule and the larger-diameter tubule is known as the B-tubule. Each outer doublet of tubules is joined to adjacent doublets by a nexin bridge. The A-tubule of each doublet is joined to the center sheath of the axoneme by a radial spoke structure. Ciliary motion: bending of a cilium involves ATP-dependent walking of motor domains of A-tubule dynein arms along adjacent B-tubules, toward the minus-end. This causes the sliding of MT doublets.

According to the structure of the dynein nanomotor, compared to kinesin and myosin, it has two potential lever arms: the MT-binding stalk and the stem. The action of these two lever arms also opens up the possibility of a longer step size than kinesin, as observed for cytoplasmic dynein [59]. It has been reported that cytoplasmic dynein molecules primarily take large steps

(24–32 nm) at low load, but advance in 8 nm steps when multiple dyneins interact with a MT and contribute to movement [59, 156]. In another study, an 8 nm step size of cytoplasmic dynein carrying peroxisomes in living cells has been reported [114]. Also it has been found that dynein moves primarily in 8 nm increments through alternating movements of its two motor domains. These findings, and the fact that the step size of the nanomotor increases at low load, suggests that the central body of dynein may function as a motorized gear box that transduces the rotation of the cargo-connected tail to the MT-connected stalk [146, 157].

A common feature of dynein nanomotors is their limping behavior. Cytoplasmic dynein appears to follow a more random path along the MT surface, frequently switching between protofilaments [150]. This nanomotor also shows frequent forth, backward motion and pauses, even when moving under no load. As a consequence, the distribution of velocities for single cytoplasmic dynein-carried beads *in vitro* is much broader than kinesin-1 [158]. For kinesin-1 driven cargo, backward motion is extremely rare. It has been suggested that the back and forth motion reflects diffusion on the MT. The ability to switch into a diffusive state where the nanomotor cannot generate force, but remains weakly bound to the MT could in principle help in avoiding traffic-jam situations. In such traffic-jam situations, oppositely directed cargos (driven by kinesin and dynein) get stuck on a MT because neither nanomotor will yield to the other [156]. To avoid this, a weaker dynein nanomotor with the ability to enter into a weakly bound state is desirable because it could "step back" and let the kinesin driven cargo pass by [159].

Another common feature of dynein nanomotors is their processivity. Cytoplasmic dynein is a weak nanomotor and might require additional factors such as dynactin, is a very large molecular complex containing at least 10 proteins [155] and has a MT-binding arm [151], to stay attached to the MT during motion and help dynein to move [151, 160]. However, dynein-based transport in the cell is robust [161, 162]. Since dynactin has a MT-binding arm, it appears that dynactin is essential for enhancing dynein's processivity [151]. *In vivo*, the loss of accessory proteins such as dynactin inhibits dynein function, and *in vitro* studies indicate that single dynein processivity is doubled by the presence of dynactin [151, 163-165]. Therefore, dynactin has been suggested to be a processivity factor for dynein, from the observations that dynactin can bind independently to MT and that dynactin increases the run-length of single dynein nanomotors in an *in vitro* assay [151]. Dynactin also links cytoplasmto dynein to its membranebased cargos in cells [48]. Because dynactin is not a motor, the most reasonable explanation is that multiple dynein molecules are bound to cellular cargo. This idea led Mallik *et al.* [156] to examine the motion of artificial cargo bound to more than one dynein molecule. They have indicated that single dynein runs were usually short, with pauses and segments of backward motion. It has been then found that for multiple dynein driven motion, the properties of motion were significantly improved and the diffusive component of motion was reduced. In fact, movement by just two dynein molecules virtually eliminates backward motions and allows cargo to run four times the distance seen with a single dynein nanomotor. In other words, the addition of a second dynein molecule to a cargo, even in the absence of dynactin, allows transport along MTs for several microns. Several previous studies examined the movement of artificial beads with different numbers of bound dynein molecules and concluded via statistical methods that a single dynein can take multiple steps per encounter with its MT track [59, 150, 151]. Thus, with multiple dyneins and dynactin present on cellular cargos, retrograde transport in cells is understandably robust. A theoretical model of dynein's function based on Monte Carlo simulations has also been developed [166]. This allows prediction of the expected force–velocity curve for dynein, reflecting how both its step size and enzymatic cycle could depend on load. Comparing the force–velocity curve to that of kinesin-1, one concludes that dynein velocity is affected more strongly than kinesin on the application of load.

3.3. FUNCTION OF THE MYOSIN PROTEIN NANOMOTOR

Myosins function in a wide variety of cellular tasks, from cellular transport to muscle contraction. Many cellular movements, such as cellular transport and muscle contraction, depend on the interactions between actin filaments and myosin [167, 168]. All characterized myosin nanomotors move towards the barbed plus-end of actin filaments with an exception, myosin-VI, that we point to it later [169]. A general mechanism of myosin motion is showed in Figure 10. As can be observed from Figure 10, it begins with myosin-ADP bound to actin and point to interaction between myosin and actin filament. Therefore the release of ADP and the binding of ATP result in the dissociation of myosin from actin. This behavior stands in contrast with the behavior of kinesin. After hydrolyzing of ATP to ADP and P_i, P_i releases and leads to a conformational change (power stroke) that increases the affinity of the myosin head for actin and allows the lever arm to move back to its initial position. After the release of P_i, the myosin remains tightly bound to the actin and the cycle can repeat again. Analysis of existing crystal structures of myosins leads to their classification into the following states: the near-rigor state is believed to be a weakly bound state that occurs shortly after detaching from actin [170]; the detached state is argued to be a stable ATP state [171] with unconstrained converter/ lever arm; the transition state is a pre-power stroke, weakly bound state that occurs after hydrolysis [172]. A communication pathway between the nucleotide binding pocket and the actin binding region is proposed: the binding with actin triggers the opening of the nucleotide binding pocket and vice versa [173]. Motivated by this dynamical linkages among the nucleotide binding site, the actin binding site and the force-generating converter/lever arm, the following power stroke mechanism was proposed: right after the hydrolysis, actin binding triggers the opening of the nucleotide binding pocket which allows the release of phosphate (and ADP at a later time) and induces a power stroke, the swinging motion of the lever arm from the up to down orientation. During this transition, myosin goes from a weakly bound transition state to a strongly bound rigor state. A reverse stroke of the lever arm from the down to up orientation is expected to occur upon ATP binding. Since the power stroke always moves the lever arm by the same angle, the length of the lever arm determines how fast the cargo will move [174]. A longer lever arm will cause the cargo to traverse a greater distance even though the lever arm undergoes the same angular displacement. Differences in myosin step sizes are not only controlled by lever arm length, but also by substantial differences in the degree of lever arm rotation [175].

Myosin nanomotors are in two groups: non-muscle myosins and muscle type myosins [85]. The non-muscle members of the myosin family (unconventional myosins) are involved in organelle transport, using actin filaments of the cytoskeleton as filaments, their mechanism of action is very similar to that described for kinesin. Moreover, in unconventional myosins the coiled-coil tail ends in cargo binding domain similar to that described for kinesin structure. As an example, myosin-V is an unconventional two-headed myosin that carries its intracellular cargo for long distances in a hand-over-hand fashion, toward the actin filament plus-end, very similar to kinesin [176]. Myosin-V is a processive actin-based nanomotor with a high duty ratio (about 1) [122]. Myosin-V and kinesin-1 show similarity of the core motor domain, suggesting that they have some similarities in function [177]. They exhibit robust unidirectional motion, with rare instances of pausing or backward motion [131, 135, 178]. The step size of myosin-V was found to be 37 nm/ATP hydrolyzed [176]. Similar to myosin-V, myosin-VI is another unconventional myosin nanomotor capable of taking multiple steps (processive movement) on an actin filament without detachment in a hand-over-hand fashion [179-181]. However, myosin-VI is a reverse direction actin-based nanomotor capable of taking large steps (30–36 nm) when dimerized and moves in opposite direction from myosin-V, toward actin filament minus-end [180, 182].

Muscle type myosin is an important component of the muscle and is the protein nanomotor that drives muscle contractions. Muscle contraction occurs when 2 sets of interdigitating filaments i.e. thin filament and thick filament, slide past each other. In order to form the thick filament, the tails of many myosins wind together into thick bipolar structures with the myosin heads protruding at both ends of a bare region in the center [183]. To achieve muscle contraction, thick filaments bind to thin filaments in an ATP-dependent fashion and the myosin heads take turns pulling on the actin filaments so that the muscle can keep shortening in a hand-over-hand pulling fashion.

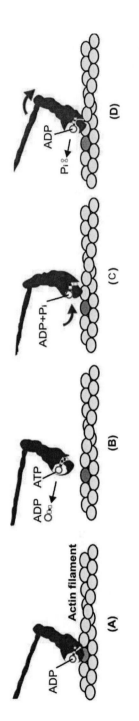

Figure 10. Schematic of interaction between myosin and actin filament: (A) For beginning stage, it is supposed that a myosin head in the ADP form is bound to an actin filament; (B) The exchange of ADP for ATP results in the release of myosin from actin and the cleft is open; (C) The ATP pocket is close and hydrolysis of ATP to ADP and P_i occurs, which remain bound to myosin. Hydrolysis of ATP to ADP and P_i allows the myosin head to rebind at a site displaced along the actin filament. This binding is weak. (D) The release of P_i leads to a conformational change that increases the affinity of the myosin head for actin, to a tight binding, and resets the orientation of the lever arm.

This process is performed multiple times until the desired muscle contraction is reached [5]. Therefore, muscle type myosins do not walk along the actin filament, but bind the substrate (actin filament) tightly, perform a conformational change that pulls the filaments and then release the substrate without interfering with other myosin molecules and without synchronization upon a single thin filament actin molecule [184]. This indicates that muscle myosins are nonprocessive nanomotors [185-188]. Muscle contraction is also regulated by Ca^{2+}-sensitive molecular switches on the myosin or actin filaments, depending on the muscle and species [189-191]. At low Ca^{2+} concentration levels, actin–myosin interaction is inhibited and actin-activated myosin ATPase is low. In this situation the muscle is at rest and tropomyosin is covered all of the myosin binding sites on the actin filament and prevents myosin from interacting with actin. At high Ca^{2+} concentrations, the inhibition of actin-myosin interaction is removed and myosin ATPase is activated by actin. Because, when a calcium ion fills the Ca^{2+} binding site of troponin, it leads to troponin acts like a lever to pull tropomyosin away from the myosin binding sites on actin and myosin could be able to interact with actin and to activate the muscle [190]. Already almost 20 years ago, researchers showed that the myosin nanomotors of skeletal muscle were also functional *in vitro*. Myosin immobilized onto a glass plate was able to move actin filaments at a velocity of about 3-4 μm per second in the presence of ATP [192]. Another evidence for function of myosin protein nanomotors of skeletal muscle *in vitro* by using optical trap has been reported [188], in which the size of the displacement steps was fairly uniform with an average size of 11 nm under conditions of low load, and single force transients averaging 3-4 pN were measured under isometric conditions. Some theoretical models for cooperative steady-state ATPase activity of myosin S1 on regulated actin has been extended [193]. It has been reported that exact solution of the general steady-state problem requires Monte Carlo calculations.

3.4. COMPARISON OF PROCESSIVITY IN
LINEAR PROTEIN NANOMOTORS

According to the above description of the functions of kinesin, dynein and myosin nanomotors, it can be concluded that each of these three classes of protein nanomotors has members with processive motion in their families. Of the three major types of these nanomotors only kinesins are strongly processive [147]. Kinesin processivity arises because the two motors of the dimeric molecule coordinate their chemical and mechanical cycles to move hand-over-hand and keep one motor bound to the MT throughout motion. By contrast, the motor domains of most dimeric myosins operate independently (a likely exception is the processive transport specialist myosin-V). Myosin heads spend most of their cycle unbound from an actin filament, therefore many myosin molecules are required to sustain filament movement in reconstitution assays. Cytoplasmic dynein appears to fall between these extremes. Reconstitution studies reveal that short (< 1μm) minus-end runs are frequently interrupted by pauses and plus-end motion. So, while dynein molecules can remain attached to MTs for a long time, they appear to be processive only part of the time [147]. However, according to some exception examples in section 3.1 for nonprocessive kinesin and in section 3.3 for processive myosin, not every member of the kinesin superfamily is processive and not every myosin is nonprocessive.

3.5. CARGO TRANSPORT BY
SEVERAL LINEAR PROTEIN NANOMOTORS

In biological cells, protein nanomotors transport various types of cargo particles such as vesicles, organelles, and filaments. This cargo transport is typically performed by several nanomotors as revealed by electron microscopy [194, 195] and single particle tracking [114, 196-198]. Therefore, in biological cells, the motion of cargo particles along MTs is often observed to be bidirectional in the sense that the particle frequently switches its direction of motion. For example, multiple dyneins and kinesins attach to, and move, single lipid droplets along MTs in bidirectional (back and forth) fashion inside the syncytial *Drosophila* embryo [161, 196]. Such a strategy seems quite widespread [30, 199-210]. Since both kinesin and dynein nanomotors are bound to cargo particles, it is rather natural to assume that the bidirectional motion arises from the competition between these two nanomotor species. Several proposals have been made for the molecular mechanisms underlying this competition [211, 212]. The presumably simplest mechanism consists of a 'tug-of-war' that there are two teams of plus and minus nanomotors that pull in opposite directions and the direction of motion is determined by the stronger team. However, since the number of nanomotors that actually pull varies with time for both nanomotor species, the weaker team may suddenly become the stronger one which reverses the direction of motion [213]. Monte Carlo method has been used to simulate the dynamics of the coordinated movement of kinesin and dynein nanomotors [214]. It is now clear that myosin, kinesin and dynein interact with each other either directly or indirectly, but it is still unknown how the nanomotors determine which cargo is to be transported and when to transport the cargo to its proper location within the cell [7]. Moreover, there are some features of cargo transportation by using several nanomotors. One of them is an effective way to increase the run length [213]. Another feature is for nonprocessive nanomotors such as muscle myosin that processivity can be increased by multiple nanomotors combining to form higher-order assemblies [215].

3.6. REGULATION AND COORDINATION OF LINEAR PROTEIN NANOMOTORS

Regulatory mechanisms must exist to coordinate the cytoskeleton-based translocation machinery with the events of vesicle assembly. Motor protein-mediated motility of organelles and transport vesicles requires spatial, temporal, and directional regulation [37]. Mechanisms for spatial regulation must exist to ensure that the nanomotors bind or are active only on the assembled vesicles and not on the donor organelle. This implies that a motor binding site is created or activated during the process of vesicle assembly. Furthermore, temporal coordination of vesicle assembly and motor-mediated motility must occur so that translocation away from the donor organelle occurs only after the completion of cargo packaging and vesicle scission. Finally, a given type of vesicle may mediate multiple trafficking steps within the cell. Thus, directional regulation may be required to select the correct protein nanomotor for each step. A number of methods for the spatial as well as temporal and directional control of filament movement have been developed. Spatial control has been achieved using topographical features [216-219], chemical surface modifications [218, 220-222], and also a combination of both [223-226]. Temporal control has been approximated by manipulating the ATP concentration [217, 227]. Directional control of MT gliding assay, that the nanomotors are immobilized on a surface and the filaments glide over the assembly, has been achieved using high enough electric fields [228-230], optical gradients produced by focusing a laser beam [231], and hydrodynamic flow fields [232, 233]. Furthermore, directional control of MT stepping assay, that the filaments are laid out on the surface where they form tracks for the nanomotors to move along, has been obtained using hydrodynamic flow fields [234-237]. A possible rationalization of the hierarchy of motor strength (kinesin-1 or kinesin- 2 > myosin-V > dynein) and processivity (kinesin-1 or kinesin-2 > dynein) as potentially useful for cooperative MT and actin filament-based motion has been reported [7]. In some systems, moreover, the same cargo can move on both MT and actin filaments, switching between nanomotors in the course of motion. Cargos moving this way include pigment granules [17, 18], axonal vesicles [238, 239], mitochondria [240] and endosomes [241, 242]. A functional collaboration [243] can then exist between MT and actin filament networks, and there have been suggestions that nanomotors associated with each network coordinate achieve the requisite subcellular distribution of cargo [26, 238]. At a global level, therefore, the intracellular transport machinery appears to regulate the relative activity of different classes of nanomotors.

On the basis of these findings, it can be concluded that all of these three protein nanomotors may generate a power stroke through related catalytic mechanisms. These protein nanomotors cycle between forms having high or low affinity for the filament tracks in response to ATP binding and hydrolysis, enabling a bind, pull, and release mechanism that generates motion. Moreover, they are often modeled with the Monte Carlo methods because the motor events are stochastic in their own nature.

APPLICATIONS OF THE LINEAR PROTEIN NANOMOTORS

Study of biological applications of nanotechnology will be important to the future of biological research, engineering and medical science [244]. Therefore, recent explosion of research in nanotechnology, combined with outstanding advances in molecular biology has created new interests in bio-nanorobots and biological nanomachines [9, 85, 245, 246]. A molecular machine has been defined as a discrete number of molecular components that have been designed to perform mechanical-like movements (output) in response to specific stimuli (input) [247]. The idea behind the biological nanomachine development is to use various biological elements as nanomachine and nanorobot components, such as nanomotors, that perform the same function in response to the same biological stimuli but in an artificial setting. To achieve this rather long-term goal, prototyping tools based on molecular dynamics (MD) simulators should be developed in order to understand the molecular mechanics of proteins and develop dynamic and kinematic models to study their performances and control aspects. A molecular mechanics study using a molecular dynamics software (NAMD) coupled to virtual reality (VR) techniques for intuitive bio-nanorobotic prototyping has been reported [248]. Their use as elementary bio-nanorobotic components were also simulated and the results have been discussed. Their objective had been to interface MD and kinematics computations with real-time truly VR simulations and measure the force, position and energy feedback for design evaluation of bio-nanorobots. Based on VR technology and MD simulators, their long-term goal is to prototype virtually bio-nanorobotic systems and control their movements in corresponding biological environment [249]. The ability to interact with a computer-generated object in the same manner that a person would interact with a physical object to investigate its structure by simply moving around it, to change its position by grabbing the object and moving the hand in space, without such artificial devices as computer mice would be the ultimate goal. It will allow the roboticians to use mechanical force to control the dynamics, time evolution and fate of chemical and biochemical reactions when connecting different bio-nanorobotic components together in series or parallel. The structural and functional analysis of biological macromolecules has reached a level of resolution that allowed mechanistic interpretations of molecular action, giving rise to the view of enzymes as molecular machines [250]. This machine analogy is not merely metaphorical, as bioanalogous nanomachines actually are being used as nanomotors in the fields of nanotechnology and robotics.

Another example of an engineering application of protein nanomotor is in molecular shuttles. Molecular shuttles have been built from protein nanomotors capable of moving cargo along engineered paths. The first prototypes of molecular shuttles are hybrid devices that employing protein nanomotors in a synthetic environment [251, 252]. Reports on the key problems for the construction of a molecular shuttle are as follow: guiding the direction of motion, controlling the speed, and loading and unloading of cargo. Various techniques, relying on surface topography and chemistry as well as flow fields and electric fields, have been developed to guide the movement of molecular shuttles on the surfaces [251]. Furthermore, the control of ATP concentration, acting as a fuel supply, can serve as a means to control the speed of movement. Finally, the loading process requires the coupling of cargo to the shuttle, ideally by a strong and specific link. Applications of molecular shuttles can be envisioned, e.g. in the field of nano-electro-mechanical systems (NEMS), where scaling laws favor active transport over fluid flow, and in the bottom-up assembly of novel materials [251]. In another first steps in the development of a tool kit to utilize protein nanomotors for the construction of nanoscale assembly lines has been realized [217]. In that research, alternative methods of controlling the direction of motion of MTs on engineered kinesin tracks, how to load cargo covalently to MTs, and how to exploit UV-induced release of caged ATP combined with enzymatic ATP degradation by hexokinase to turn the shuttles on and off sequentially have been illustrated.

A long-term goal of nanobiotechnology is to build tiny devices that respond to the environment, perform computations and carry out tasks. A nanodevice is a tiny entity, a gadget or machine, capable of performing a task. Nanodevices might respond to the environment through proteins with built-in switches that operate in a simple on/off way or through more finely tuned and complex logic gates with graded or multiple inputs. In this way, nanodevices will sense their environment. Nanodevices might use protein nanomotors to move linearly, by rotation, or in a more complex three-dimensional manner. More advanced functions might include transport (uptake, movement and delivery of cargos utilizing protein transporters and pores) and chemical transformation, by enzymatic catalysis, for example. To perform these functions, the nanodevice must use energy and might even transduce and store it by using, for example, the biological energy currency of ATP. This description proposes that protein nanomotors can be function as components for nanodevices [253]. Some speculations about what might be possible and examine the progress that has been made in making components for nanodevices has been reported including that three classes of protein components for nanodevices are presented in order of complexity: planar crystalline arrays, engineered protein pores, and protein nanomotors [253]. Remarkably, protein nanomotors differ fundamentally from artificial devices in that the conversion from chemical energy to mechanical energy is done directly, rather than via an intermediary stage, as in, e.g., heat in thermal engines. This fundamental difference, which translates into a very high energy efficiency of these natural devices compared to artificial mechanical devices, together with their very small scale, prompted an increasing number of studies focused on the integration of protein nanomotors in hybrid micro- and nanodevices [254]. Considerable progress has been made in building protein components for such devices [255].

It has been reported that biological nanomotors, in particular protein nanomotors, are ideally suited to introduce chemically powered movement of selected components into devices engineered at the micro- and nanoscale level [256]. It was showed the design of such hybrid bio/nano-devices requires suitable synthetic environments, and the identification of unique applications. Protein nanomotors also have been integrated in the last decade in primitive nanodevices based on the motility of nano-biological objects in micro- and nano-fabricated structures. However, the motility of microorganisms powered by protein nanomotors has not been similarly exploited. Biocomputation with motile biological agents is in its infancy, but the development challenges appear to be more related to design and operation rather than fabrication. According to this subject, application of protein nanomotors in micro- and nano-biocomputation devices has been reported [254]. It was proposed that if it is conceived the use of 'non-programmable' motile biological agents, such as protein nanomotors, in nano-fabricated networks that code mathematical problems, then these will also solve the encoded problems, provided that the lack of 'self-program-ability' is replaced by means of control (and record) the movement of the nano-agents in the network. Cells perform computations when they explore (at times limited and confined) available space for nutrients and to allow for their growth. In that work the space searching of cells was in a sense a spatial biocomputation that was scaled down from multicellular 'intelligent' organisms, e.g., mice, octopi and humans solving mazes. Conversely, if one can purposefully code mathematical problems in (micro) fabricated networks and let 'self-programmable' biological agents, such as microorganisms, explore this network, then this space search will solve the mathematical problem encoded. They solved mazes using fungi-microorganisms that are perfectly adapted to negotiate narrow natural networks, e.g., cracks in rocks, in order to find nutrients often in nutrient-scarce environments. They chose fungi because protein nanomotors are also critical to fungal growth.

Perhaps the most exciting goal of biological nanomotors is the molecular repair of the human body. Medical nanorobots are envisioned that could destroy viruses and cancer cells, repair damaged structures, remove accumulated wastes from the brain, bring the body back to a state of youthful health and deliver drugs in body [257]. Recently, researchers have begun to shift their efforts towards developing applications that utilize this technology in novel ways, such as part of nanomachines [250, 258] and for the delivery of genes or drugs to the nucleus of cells or to the central nervous system. On the other hand, complex biological environments can pose significant barriers to efficient therapeutic drug and gene delivery [259]. There has been little effort to overcome the barrier of transporting DNA towards the nucleus despite evidence that the mobility of DNA in the cytosol might be a barrier to gene transfer. Cohen *et al.* [260] attempted to use protein nanomotors as a DNA delivery vehicle by two approaches (Figure 11). The first approach involves creating a fusion protein that contains a minus-end microtubular motor domain and a DNA-binding domain from GAL4, a yeast transcription factor (Figure 11A). An alternative gene delivery method that have been proposed was the use of dynein protein nanomotor and a biomolecular adaptor for retrograde transport (BART), a synthetic adaptor that links DNA or other novel cargo to dynein (Figure 11B). In another work, Gunawardena *et al.* [261] reported that protein nanomotors might also be applied as a drug delivery vehicle to the cell bodies of motor neurons by axonal transport. They concentrated on MT-based protein nanomotors, their linkers, and cargos and discussed how factors in the axonal transport pathway contribute to disease states. They reported, as additional cargo complexes and transport pathways are identified, an understanding of the role these pathways play in the development of human disease will hopefully lead to new diagnostic and treatment strategies. It has been also proposed that traffic jams of protein nanomotors are involved in a variety of neurodegenerative diseases [262, 263]. Each cell of our body contains a huge number of small vesicles which exhibit complex patterns of intracellular traffic: some vesicles travel from the cell center to the periphery and vice versa, some shuttle between different organelles or cellular compartments. All of this traffic is based on two molecular components: cytoskeletal filaments and protein nanomotors. As one further increases the nanomotor concentrations, the filaments start to become overcrowded and the nanomotor flux becomes reduced by traffic jams [264-266]. Lipowsky *et al.* [213] proposed that, it is necessary to understand these traffic phenomena in a quantitative manner because active biomimetic systems based on these nanomotors and filaments have many potential applications in bioengineering, pharmacology and medicine. Such applications include sorting devices for biomolecules, motile drug delivery systems, molecular shuttles in 'labs-on-a-chip', and switchable scaffolds for tissue engineering. These results indicate that micro- and nanotechnologies are enabling the design of novel methods and materials in the application of micro- and nanosystems for drug administration.

Another potential application of protein nanomotors is reported by Bunk *et al.* [267]. They proposed that they reconstructed *in vitro* the behavior of two nanomotor proteins, myosin and actin, responsible for the mechanical action of the muscle cells. By transferring this *in vivo* system to an artificial environment, they were able to study the interaction between the proteins in more detail, as well as investigating the central mechanism of force production. Thus, the creation of *in vitro* nanostructured, ordered interactions between actin and myosin is of interest for potential applications in nanotechnology, such as the development of a 'factory on a chip'. Biological nanomotors, specialized proteins, also have unique advantages for integrated nanomaterials and systems. The integration of protein nanomotors into an artificial environment enables us to explore new space in the development of nanotechnology. Potential applications of devices and materials integrating nanomotors abound in biosensing, nanofluidics, molecular electronics, digital light processing, nanoscale and macroscale actuation, and adaptive materials. Some of these applications derive their inspiration from long-replaced macroscale technologies [268].

On the basis of these findings, it can be concluded that there are many challenges in applying protein nanomotors to nanotechnology. All these developments will be solely up to the imagination and skills of researchers, both in engineering and in natural sciences to envision the future and limitations of use for such highly intriguing devices that Nature has designed and perfected over billions of years and is giving to us for our own use. Moreover, it can be noted that these nanomotors have some advantages and disadvantages that will be contributing in their future applications [9]. Thus, in termination of this section we point to these advantages and disadvantages gradually. The potential advantages in developing of bionanomotors are: (1)

high efficiency that is but one feature making protein nanomotors attractive for nanotechnological applications; (2) ease of cheaply manufacture in vast quantities; (3) extent availability [269]; (4) small size so they can operate in a highly parallel manner; (5) easy to produce so they can be modified through genetic engineering; a wide array of biochemical tools have been developed to manipulate these proteins outside the cell [9]. Moreover, according to the biological processes of the protein nanomotors, these processes have received increasing interest owing to their cost, effectiveness and environmental benignity [270, 271]. Two main disadvantages of protein nanomotors are their limited *in vitro* lifetime, and the narrow range of environmental conditions that they are able to tolerate [267]. It is interesting to speculate about the limits of extending the lifetime of biological nanomachines [272], the causes of failure that cannot be avoided by changes in the environment, and the tradeoffs involved in balancing lifetime, force output, speed, and efficiency when designing high-performance synthetic nanomotors [273].

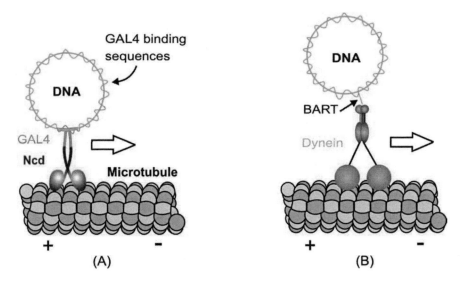

Figure 11. Strategies to use protein nanomotors to actively transport DNA to the nucleus. (A) The chimeric protein was consisted of a *ncd* and a DNA-binding domain, such as that from GAL4, a transcription factor from yeast. (B) A biomolecular adaptor for retrograde transport (BART) was linked DNA to endogenous dynein which then carried DNA along MTs. BART was covalently linked to plasmid DNA.

CONCLUSIONS

Nanotechnology would be the key technology of the current century. As new technologies and methods are developed, it is expected that knowledge of naturally occurring biological nanomotors will be helpful in constructing synthetic counterparts and therefore nanorobots. The most complex biological nanomotors are found in within cells are far more complex than any nanomotors that have yet been artificially constructed. These include protein nanomotors, such as, kinesin that moves cargo inside cells away from the nucleus along MTs, dynein that produces the axonemal beating of cilia and flagella as well as transports cargo along MTs towards the cell nucleus, and myosin that is responsible for muscle contraction as well as cellular transportation along actin filaments. The majority of active transport in the cell is driven by these three classes of protein nanomotors. For short distance transport, myosin carries its cargo along short actin filaments. For long distance transport, kinesins and cytoplasmic dynein carry cargo along MTs. Thus, protein nanomotors are biological nanomachines acting as the essential agents of movement in living organisms and are responsible for the organism activity. These tiny sized macromolecules generate different motions of the body and also the subject of interest for nanotechnology. Recent advances in understanding how protein nanomotors work, has raised the possibility that they might find applications as protein-based nanomachines and nanorobots. Therefore, promoted understanding and engineering of these nanomotors might lead to applications and solutions to problems beyond those that are only imagined today. The horizons are broad and the ability to engineer these systems might soon create fact from fantasy.

REFERENCES

[1]   Feynman, RP. Engineering and Science Magazine of Cal. *Inst. of Tech.*, 1960, 23, 22-36.
[2]   Taniguchi, N. Proc. Intl. Conf. Prod. *Eng. Tokyo.*, 1974, 5-10.
[3]   Khataee, HR; Khataee, AR. Second Student *Congress of Recent Advances in Chemistry.*, 2008, 1.
[4]   Mansoori, GA; Suwono, A. the 5th International *Conference on Fluid and Thermal Energy Conversion.*, 2006, 10-14.
[5]   Vogel, PD. Eur. *J. Pharm. Biopharm.*, 2005, 60, 267-277.
[6]   Jamali, Y; Lohrasebi, A; Rafii-Tabar, H. *Physica A.*, 2007, 381, 239-254.
[7]   Mallik, R; Gross, SP. *Curr. Biol.*, 2004, 14, R971-R982.
[8]   Mallik, R; Gross, SP. *Physica A.*, 2006, 372, 65-69.
[9]   Khataee, HR; Khataee, AR. *Nano*, 2009, 4 (2), 55-67.
[10] Rayment, I; Holden, HM. Trends. *Biochem. Sci.*, 1994, 19, 129-134.
[11] Niemeyer, CM; Mirkin, CA. Nanobiotechnology: Concepts, Applications and Perspectives; Wiley-VCH: *Weinheim*, 2004, Vol. 1, 185-200.
[12] Downing, KH; Nogales, E. Curr. *Opin. Cell. Biol.*, 1998, 10, 16-22.
[13] Amos, LA; Hirose, K. Curr. Opin. *Cell. Biol.*, 1997, 9, 4-11.
[14] Wade, RH; Hyman, AA. Curr. Opin. *Cell. Biol.*, 1997, 9, 12-17.
[15] Howard, J; Hyman, AA. *Nature.*, 2003, 422, 753-758.
[16] Rodionov, V; Nadezhdina, E; Borisy, G. Proc. Natl. Acad. Sci. USA. 1999, 96, 115-120.
[17] Rodionov, V; Yi, J; Kashina, A; Oladipo, A; Gross, SP. *Curr. Biol.*, 2003, 13, 1837-1847.
[18] Gross, SP; Tuma, MC; Deacon, SW; Serpinskaya, AS; Reilein, AR; Gelfand, VI.J. *Cell. Biol.*, 2002, 156, 855-865.
[19] Burakov, A; Nadezhdina, E; Slepchenko, B; Rodionov, V.J. Cell. Biol. 2003, 162, 963-969.
[20] Goldstein, LS; Yang, Z. Annu. *Rev. Neurosci.*, 2000, 23, 39-71.
[21] Musch, A. *Traffic.*, 2004, 5, 1-9.

[22] Hartwig, JH; Bokoch, GM; Carpenter, CL; Janmey, PA; Taylor, LA; Toker, A; Stossel, TP. *Cell.*, 1995, 82, 643-653.

[23] Steinmetza, MO; Stofflera, D; Aebi, AHAB. *J. Struct. Biol.*, 1997, 119, 295-320.

[24] Fath, KR; Lasek, RJ. *J. Cell. Biol.*, 1988, 107, 613-621.

[25] Bray, D; Bunge, MB. *J. Neurocytol.*, 1981, 10, 589-605.

[26] Brown, SS. Annu. Rev. Cell. Dev. *Biol.*, 1999, 15, 63-80.

[27] Lewis, AK; Bridgman, PC. *J. Cell. Biol.*, 1992, 119, 1219-1243.

[28] Svitkina, TM; Verkhovsky, AB; McQuade, KM; Borisy, GG. *J. Cell. Biol.*, 1997, 139, 397-415.

[29] Snider, J; Lin, F; Zahedi, N; Rodionov, V; Yu, CC; Gross, SP. Proc. Natl. Acad. *Sci. USA.*, 2004, 101, 13204-13209.

[30] Nascimento, AA; Roland, JT; Gelfand, VI. Annu. Rev. *Cell. Dev. Biol.*, 2003, 19, 469-491.

[31] Brady, S. *T. Nature.*, 1985, 317, 73-75.

[32] Vale, RD; Reese, TS; Scheetz, MP. *Cell.*, 1985, 42, 39-50.

[33] Allen, RD; Metuzals, J; Tasaki, I; Brady, ST; Gilbert, SP. *Science.*, 1982, 218, 1127-1129.

[34] DeBoer, SR; You, Y; Szodorai, A; Kaminska, A; Pigino, G; Nwabuisi, E; Wang, B; Estrada-Hernandez, T; Kins, S; Brady, ST; Morfini, G. *Biochemistry-US.*, 2008, 47, 4535-4543.

[35] Yang, JT; Laymon, RA; Goldstein, LS. *B. Cell.*, 1989, 56, 879-889.

[36] Hirokawa, N; Pfister, KK; Yorifuji, H; Wagner, MC; Brady, ST; Bloom, G. *S. Cell.*, 1989, 56, 867-878.

[37] Hehnly, H; Stamnes, M. *FEBS. Lett.*, 2007, 581, 2112-2118.

[38] Vale, RD. *J. Cell. Biol.*, 1996, 135, 291-302.

[39] Woehlke, G; Ruby, AK; Hart, CL; Ly, B; Hom-Booher, N; Vale, R. *D. Cell.*, 1997, 90, 207-216.

[40] Rice, S; Lin, AW; Safer, D; Hart, CL; Naber, N; Carragher, BO; Cain, SM; Pechatnikova, E; Wilson-Kubalek, EM; Whittaker, M; Pate, E; Cooke, R; Taylor, EW; Milligan, RA; Vale, RD. *Nature.*, 1999, 402, 778-784.

[41] Case, RB; Rice, S; Hart, CL; Ly, B; Vale, R. *D. Curr. Biol.*, 2000, 10, 157-160.

[42] Kozielski, F; Sack, S; Marx, A; Thormahlen, M; Schonbrunn, E; Biou, V; Thompson, A; Mandelkow, EM; Mandelkow, E. *Cell.*, 1997, 91, 985-994.

[43] Hackney, DD; Levitt, JD; Suhan, J. *J. Biol. Chem.*, 1992, 267, 8696-8701.

[44] Friedman, DS; Vale, RD. Nat. *Cell. Biol.*, 1999, 1, 293-297.

[45] Seiler, S; Kirchner, J; Horn, C; Kallipolitou, A; Woehlke, G; Schliwa, M. Nat. *Cell. Biol.*, 2000, 2, 333-338.

[46] Vallee, RB; Williams, JC; Varma, D; Barnhart, LE. *J. Neurobiol.*, 2004, 58, 189-200.

[47] King, SM. *J. Cell. Sci.*, 2000, 113, 2521-2526.

[48] Hirokawa, N. *Science.*, 1998, 279, 519-526.

[49] King, SM. *Cell. Biol. Int.*, 2003, 27, 213-215.

[50] Neuwald, AF; Aravind, L; Spouge, JL; Koonin, EV. *Genome. Res.*, 1999, 9, 27-43.

[51] Mocz, G; Gibbons, IR. *Biochemistry-US.*, 1996, 35, 9204-9211.

[52] Mocz, G; Helms, MK; Jameson, DM; Gibbons, IR. *Biochemistry-US.*, 1998, 37, 9862-9869.

[53] Gibbons, IR; Gibbons, BH; Mocz, G; Asai, D. *J. Nature.*, 1991, 352, 640-643.

[54] Samso, M; Koonce, MP. *J. Mol. Biol.*, 2004, 340, 1059-1072.

[55] Reck-Peterson, S. L; Vale, RD. Proc. Natl. *Acad. Sci. USA.*, 2004, 101, 1491-1495.

[56] Takahashi, Y; Edamatsu, M; Toyoshima, YY. Proc. Natl. *Acad. Sci. USA.*, 2004, 101, 12865-12869.

[57] Mocz, G; Gibbons, IR. *Structure.*, 2001, 9, 93-103.

[58] Gibbons, IR; Lee-Eiford, A; Mocz, G; Phillipson, CA; Tang, WJ; Gibbons, BH. *J. Biol. Chem.*, 1987, 262, 2780-2786.

[59] Mallik, R; Carter, BC; Lex, SA; King, SJ; Gross, SP. *Nature.*, 2004, 427, 649-652.

[60] Burgess, SA; Walker, ML; Sakakibara, H; Knight, PJ; Oiwa, K. *Nature.*, 2003, 421, 715-718.

[61] Schliwa, M; Woehlke, G. *Nature.*, 2003, 422, 759-765.

[62] **Koonce, MP; Samsó, M.** Trends. *Cell. Biol.*, 2004, 14, 612-619.

[63] Sakato, M; King, SM. *J. Struct. Biol.*, 2004, 146, 58-71.

[64] Kon, T; Mogami, T; Ohkura, R; Nishiura, M; Sutoh, K. Nat. *Struct. Mol. Biol.*, 2005, 12, 513-519.

[65] Marchese-Ragona, SP; Johnson, KA. Structure and Subunit Organization of the Soluble Dynein Particle; In Cell Movement: *Alan R. Liss*, New York, 1989, Vol. 1, 37-48.

[66] Goodenough, UW; Heuser, JE. Structure of the Soluble and in situ Ciliary Dyneins Visualized by Quick-Freeze Deep-etch Microscopy; In Cell Movement: *Alan R. Liss*, New York, 1989, Vol. 1, 121-140.

[67] Gee, MA; Heuser, JE; Vallee, RB. *Nature.*, 1997, 390, 636-639.

[68] Goodenough, U; Heuser, J. *J. Mol. Biol.*, 1984, 180, 1083-1118.

[69] Koonce, MP. *J. Biol. Chem.*, 1997, 272, 19714-19718.

[70] Lowey, S; Slayter, HS; Weeds, AG; Baker, H. *J. Mol. Biol.*, 1969, 42, 1-20.

[71] Cope, MJT; Whisstock, J; Rayment, I; Kendrick-Jones, J. Structure., 1996, 4, 969-987.

[72] Rayment, I. Structure., 1996, 4, 501-504.

[73] Rayment, I; Rypniewski, WR; Schmidt-Base, K; Smith, R; Tomchick, DR; Benning, MM; Winkelmann, DA; Wesenberg, G; Holden, H. M. Science., 1993, 261, 50-58.

[74] Fisher, AJ; Smith, CA; Thoden, J; Smith, R; Sutoh, K; Holden, HM; Rayment, I. Biochemistry-US., 1995, 34, 8960- 8972.

[75] Dominguez, R; Freyzon, Y; Trybus, KM; Cohen, C. Cell., 1998, 94, 559-571.

[76] Houdusse, A; Kalabokis, VN; Himmel, D; Szent-Gyorgyi, AG; Cohen, C. Cell., 1999, 97, 459-470.

[77] Whittaker, M; Wilson-Kubalek, EM; Smith, JE; Faust, L; Milligan, RA; Sweeney, HL. Nature, 1995, 378, 748-751.

[78] Holmes, KC; Geeves, MA. Annu. Rev. Biochem., 1999, 68, 687-728.

[79] Vale, RD; Case, R; Sablin, E; Hart, C; Fletterick, R; Phil. Trans, R. Soc. Lond. B Biol. Sci., 2000, 355, 449-457.

[80] Alberts, B. cell., 1998, 92, 291-294.

[81] Watanabe, TM; Sato, T; Gonda, K; Higuchi, H. Biochem. Biophys. Res. Commun. 2007, 359, 1-7.

[82] Paschal, BM; Vallee, R. B. Nature., 1987, 330, 181-183.

[83] Vale, RD; Schnapp, BJ; Mitschison, T; Steuer, E; Reese, TS; Sheetz, MP. Cell., 1985, 43, 623-632.

[84] Vale, RD. Cell., 2003, 112, 467-480.

[85] Khataee, HR; Khataee, AR. Digest Journal of Nanomaterials and Biostructures, 2009, 4, 613-621.

[86] Vale, RD; Milligan, RA. Science., 2000, 288, 88-95.

[87] Ray, K. Physica A., 2006, 372, 52-64.

[88] Stock, MF; Guerrero, J; Cobb, B; Eggers, CT; Huang, TG; Li, X; Hackney, DDJ. Biol. Chem., 1999, 274, 14617-14623.

[89] Hackney, DD; Stock, MF. Nat. Cell. Biol., 2000, 2, 257-260.

[90] Coy, DL; Hancock, WO; Wagenbach, M; Howard, J. Nat. Cell. Biol., 1999, 1, 288-292.

[91] Kirchner, J; Seiler, S; Fuchs, S; Schliwa, M. EMBO. J., 1999, 18, 4404-4413.

[92] Yonekura, H; Nomura, A; Ozawa, H; Tatsu, Y; Yumoto, N; Uyeda, TQP. Biochem. Biophys. Res. Commun., 2006, 343, 420-427.

[93] Woehlke, G. FEBS. Lett., 2001, 508, 291-294.

[94] Woehlke, G; Schliwa, M. Nat. Rev. Mol. Cell. Biol., 2000, 1, 50-58.

[95] Walker, RA; Salmon, ED; Endow, SA. *Nature.*, 1990, 347, 780-782.

[96] McDonald, HB; Stewart, RJ; Goldstein, LS. *Cell.*, 1990, 63, 1159-1165.

[97] Howard, J; Hudspeth, AJ; Vale, RD. *Nature.*, 1989, 342, 154-158.

[98] Block, SMJ. *Cell. Biol.*, 1998, 140, 1281-1284.

[99] Cross, RA. *Trends. Biochem. Sci.*, 2004, 29, 301-309.

[100] Asbury, CL. Curr. *Opin. Cell. Biol.*, 2005, 17, 89-97.

[101] Yildiz, A; Selvin, PR. Trends. *Cell. Biol.*, 2005, 15, 112-120.

[102] Peskin, CS; Oster, G. *Biophys. J.*, 1995, 68, 202s-210s.

[103] Jülicher, F; Ajdari, A; Prost, J. *Rev. Mod. Phys.*, 1997, 69, 1269-1281.

[104] Mandelkow, E; Johnson, KA. Trends. *Biochem. Sci.*, 1998, 23, 429-433.

[105] Gilbert, SP; Moyer, ML; Johnson, KA. *Biochemistry-US.*, 1998, 37, 792-799.

[106] Astumian, RD; Derenyi, I. *Biophys. J.*, 1999, 77, 993-1002.

[107] Hancock, W; Howard, J. Proc. Natl. *Acad. Sci. USA.*, 1999, 96, 13147-13152.

[108] Fox, RF; Choi, MH. *Phys. Rev. E.*, 2001, 63, 051901.

[109] Thomas, N; Imafuku, Y; Kamiya, T; Tawada, K. Proc. Roy. Soc. *Lond. Ser. B.*, 2002, 269, 2363-2371.

[110] Hua, W; Chung, J; Gelles, J. *Science.*, 2002, 295, 844-848.

[111] Rosenfeld, SS; Fordyce, PM; Jefferson, GM; King, PH; Block, SM. *J. Biol. Chem.*, 2003, 278, 18550-18556.

[112] Klumpp, LM; Hoenger, A; Gilbert, SP. Proc. Natl. *Acad. Sci., USA.*, 2004, 101, 3444-3449.

[113] Visscher, K; Schnitzer, MJ; Block, SM. *Nature.*, 1999, 400, 184-189.

[114] Kural, C; Kim, H; Syed, S; Goshima, G; Gelfand, VI; Selvin, PR. *Science.*, 2005, 308, 1469-1472.

[115] Coppin, CM; Finer, JT; Spudich, JA; Vale, RD. Proc. *Natl. Acad. Sci., USA.*, 1996, 93, 1913-1917.

[116] Hackney, DD. *J. Biol. Chem.*, 1994, 269, 16508-16511.

[117] Gilbert, SP; Webb, MR; Brune, M; Johnson, KA. *Nature.*, 1995, 373, 671-676.

[118] Hackney, DD. *Nature.*, 1995, 377, 448-450.

[119] Block, SM; Goldstein, LS; Schnapp, B. *J. Nature.*, 1990, 348, 348-352.

[120] Svoboda, K; Block, SM. *Cell.*, 1994, 77, 773-784.

[121] Zhang, Y; Hancock, WO. *Biophys. J.*, 2004, 87, 1795-1804.

[122] Mehta, AD; Rock, RS; Rief, M; Spudich, JA; Mooseker, MS; Cheney, RE. *Nature.*, 1999, 400, 590-593.

[123] Kaseda, K; Higuchi, H; Hirose, K. Proc. *Natl. Acad. Sci., USA.*, 2002, 99, 16058-16063.

[124] Thorn, KS; Ubersax, JA; Vale, RD. J. *Cell. Biol.*, 2000, 151, 1093-1100.

[125] deCastro, MJ; Ho, CH; Stewart, RJ. *Biochemistry-US.*, 1999, 38, 5076-5081.

[126] Foster, KA; Gilbert, SP. *Biochemistry-US.*, 2000, 39, 1784-1791.

[127] Asbury, CL; Fehr, AN; Block, SM. *Science.*, 2003, 302, 2130-2134.

[128] Howard, J. Annu. Rev. *Physiol.*, 1996, 58, 703-729.

[129] Cross, RA. *Curr. Biol.*, 2004, 14, R158-R159.

[130] Svoboda, K; Schmidt, CF; Schnapp, BJ; Block, SM. *Nature.*, 1993, 365, 721-727.

[131] Coppin, CM; Pierce, DW; Hsu, L; Vale, RD. Proc. Natl. *Acad. Sci.*, USA., 1997, 94, 8539-8544.

[132] Schnitzer, MJ; Block, SM. *Nature.*, 1997, 388, 386-390.

[133] Kojima, H; Muto, E; Higuchi, H; Yanagida, T. *Biophys. J.*, 1997, 73, 2012-2022.

[134] Schnitzer, MJ; Visscher, K; Block, SM. *Nat. Cell. Biol.*, 2000, 2, 718-723.

[135] Nishiyama, M; Higuchi, H; Yanagida, T. *Nat. Cell. Biol.*, 2002, 4, 790-797.

[136] Uemura, S; Ishiwata, S. *Nat. Struct. Biol.*, 2003, 4, 308-311.

[137] Block, SM; Asbury, CL; Shaevitz, JW; Lang, MJ. Proc. *Natl. Acad. Sci. USA.*, 2003, 100, 2351-2356.

[138] Kaseda, K; Higuchi, H; Horose, K. Nat. *Cell. Biol.*, 2003, 5, 1079-1082.

[139] Higuchi, H; Bronner, CE; Park, HW; Endow, SA. *EMBO. J.*, 2004, 23, 2993-2999.

[140] Carter, NJ; Cross, R. A. *Nature.*, 2005, 435, 308-312.

[141] Taniguchi, Y; Nishiyama, M; Ishii, Y; Yanagida, T. *Nat. Chem. Biol.*, 2005, 1, 342-347.

[142] Hong, W; Shuo-Xing, D; Peng-Ye, W. *Chinese. Phys. Lett.*, 2005, 22, 2980-2982.

[143] Goedecke, DM; Elston, TC. *J. Theor. Biol.*, 2005, 232, 27-39.

[144] Burgess, SA; Walker, ML; Sakakibara, H; Oiwa, K; Knight, PJ. *J. Struct. Biol.*, 2004, 146, 205-216.

[145] Burgess, SA; Knight, PJ. *Curr. Opin. Struct. Biol.*, 2004, 14, 138-146.

[146] Reck-Peterson, SL; Yildiz, A; Carter, AP; Gennerich, A; Zhang, N; Vale, RD. *Cell.*, 2006, 126, 335-348.

[147] McGrath, JL. *Curr. Biol.*, 2005, 15, R970-R972.

[148] Karp, G. Cell and Molecular Biology: Concepts and Experiments; 4th edition; *John Wiley and Sons: Hoboken, NJ*, 2005, Vol. 1, 346-358.

[149] Paschal, BM; Shpetner, HS; Vallee, RB. J. *Cell. Biol.*, 1987, 105, 1273-1282.

[150] Wang, Z; Khan, S; Sheetz, MP. *Biophys. J.*, 1995, 69, 2011-2023.

[151] King, SJ; Schroer, TA. Nat. *Cell. Biol.*, 2000, 2, 20-24.

[152] Grissom, PM; Vaisberg, EA; McIntosh, JR. *Mol. Biol. Cell.*, 2002, 13, 817-829.

[153] Mikami, A; Tynan, SH; Hama, T; Luby-Phelps, K; Saito, T; Crandall, JE; Besharse, JC; Vallee, RBJ. *Cell. Sci.*, 2002, 115, 4801-4808.

[154] Stephens, RE. *Biochemistry-US.*, 1977, 16, 2047-2058.

[155] Ogawa, K; Inaba, K. Biochem. *Biophys. Res. Commun.*, 2006, 343, 385-390.

[156] Mallik, R; Petrov, D; Lex, SA; King, SJ; Gross, SP. *Curr. Biol.*, 2005, 15, 2075-2085.

[157] Toba, S; Watanabe, TM; Yamaguchi-Oki-moto, L; Toyoshima, YY; Higuchi, H. Proc. *Natl. Acad. Sci.*, USA., 2006, 103, 5741- 5745.

[158] Wang, Z; Sheetz, MP. *Biophys. J.*, 2000, 78, 1955-1964.

[159] Hafezparast, M; Klocke, R; Ruhrberg, C; Marquardt, A; Ahmad- Annuar, A; Bowen, S; Lalli, G; Witherden, AS; Hummerich, H; Nicholson, S; et al. *Mutations in dynein link motor neuron degeneration to defects in retrograde transport. Science.*, 2003, 300, 808-812.

[160] Schroer, TA. Annu. Rev. *Cell. Dev. Biol.*, 2004, 20, 759-779.

[161] Welte, MA; Gross, SP; Postner, M; Block, SM; Wieschaus, EF. *Cell.*, 1998, 92, 547-557.

[162] Ma, S; Chisholm, RL. *J. Cell. Sci.*, 2002, 115, 1453-1460.

[163] Roghi, C; Allan, VJ. *J. Cell. Sci.*, 1999, 112, 4673-4685.

[164] King, SJ; Brown, CL; Maier, KC; Quintyne, NJ; Schroer, TA. *Mol. Biol. Cell.*, 2003, 14, 5089-5097.

[165] Gill, SR; Schroer, TA; Szilak, I; Steuer, ER; Sheetz, MP; Cleveland, DW. *J. Cell. Biol.*, 1991, 115, 1639-1650.

[166] Singh, MP; Mallik, R; Gross, SP; Yu, CC. Proc. *Natl. Acad. Sci.*, USA., 2005, 102, 12059-12064.

[167] Brown, ME; Bridgman, PC. *J. Neurobiol.*, 2004, 58, 118-130.

[168] Arner, A; Lofgren, M; Morano, IJ. Muscle. Res. *Cell. Motil.*, 2003, 24, 165-173.

[169] Sellers, JR; Goodson, HV. *Protein Profile.*, 1995, 2, 1323-1423.

[170] Gulick, AM; Bauer, CB; Thoden, JB; Rayment, I. *Biochemistry-US.*, 1997, 36, 11619-11628.

[171] Houdusse, A; Szent-Gyorgyi, AG; Cohen, C. Proc. Natl. *Acad. Sci.*, USA., 2000, 97, 11238-11243.

[172] Smith, CA; Rayment, I. *Biochemistry-US.*, 1996, 35, 5404-5417.

[173] Zheng, W; Brooks, B. *J. Mol. Biol.*, 2005, 346, 745-759.

[174] Geeves, MA. *Nature.*, 2002, 415, 129-131.

[175] Köhler, D; Ruff, C; Meyhöfer, E; Bähler, MJ. *Cell. Biol.*, 2003, 161, 237-241.

[176] Yildiz, A; Forkey, JN; McKinney, SA; Ha, T; Goldman, YE; Selvin, PR. *Science*, 2003, 300, 2061-2065.

[177] Kull, FJ; Sablin, EP; Lau, R; Fletterick, RJ; Vale, RD. *Nature.*, 1996, 380, 550-555.

[178] Rief, M; Rock, RS; Mehta, AD; Mooseker, MS; Cheney, RE; Spudich, JA. Proc. *Natl. Acad. Sci.*, USA., 2000, 97, 9482-9486.

[179] Kellerman, KA; Miller, KG. *J. Cell. Biol.*, 1992, 119, 823-834.

[180] Wells, AL; Lin, AW; Chen, LQ; Safer, D; Cain, SM; Hasson, T; Carragher, BO; Milligan, RA; Sweeney, HL. *Nature.*, 1999, 401, 505-508.

[181] Ökten, Z; Churchman, LS; Rock, RS; Spudich, JA. Nature. Struct. *Molecular Biol.*, 2004, 11, 884-887.

[182] Park, H; Ramamurthy, B; Travaglia, M; Safer, D; Chen, LQ; Armstrong, CF; Selvin, PR; Sweeney, HL. *Mol. Cell.*, 2006, 21, 331-336.

[183] Miroshnichenko, NS; Balanuk, IV; Nozdrenko, DN. *Cell. Biol. Int.*, 2000, 24, 327-333.

[184] Ikebe, M. Biochem. Biophys. Res. *Commun.*, 2008, 369, 157-164.

[185] Reconditi, M; Linari, M; Lucii, L; Stewart, A; Sun, Y; Boesecke, P; Narayanan, T; Fischetti, RF; Irving, T; Piazzesi, G; Irving, M; Lombardi, V. *Nature.*, 2004, 428, 578-581.

[186] Piazzesi, G; Reconditi, M; Linari, M; Lucii, L; Sun, YB; Narayanan, T; Boesecke, P; Lombardi, V; Irving, M. *Nature.*, 2002, 415, 659-662.

[187] Spudich, JA; Finer, J; Simmons, B; Ruppel, K; Patterson, B; Uyeda, T. Cold. Spring. Harb. Symp. *Quant. Biol.*, 1995, 60, 783-791.

[188] Finer, JT; Simmons, RM; Spudich, JA. *Nature.*, 1994, 368, 113-119.

[189] Lehman, W; Hatch, V; Korman, V; Rosol, M; Thomas, L; Maytum, R; Geeves, MA; Van Eyk, JE; Tobacman, LS; Craig, RJ. *Mol. Biol.*, 2000, 302, 593-606.

[190] Zhao, F; Craig, RJ. *Mol. Biol.*, 2003, 327, 145-158.

[191] Lehman, W; Szent-Györgyi, AG. *J. Gen. Physiol.*, 1975, 66, 1-30.

[192] Kron, SJ; Spudich, JA. Proc. Natl. *Acad. Sci.*, USA., 1986, 83, 6272-6276.

[193] Hill, TL; Eisenberg, E; Chalovich, JM. *Biophys. J.*, 1981, 35, 99-112.

[194] Miller, RH; Lasek, RJ. *J. Cell. Biol.*, 1985, 101, 2181-2193.

[195] Ashkin, A; Schütze, K; Dziedzic, JM; Euteneuer, U; Schliwa, M. *Nature.*, 1990, 348, 346-348.

[196] Gross, SP; Welte, MA; Block, SM; Wieschaus, EF. *J. Cell. Biol.*, 2002, 156, 715-724.

[197] Hill, DB; Plaza, MJ; Bonin, K; Holzwarth, G. *J. Eur. Biophys.*, 2004, 33, 623-632.

[198] Levi, V; Serpinskaya, AS; Gratton, E; Gelfand, V. *Biophys. J.*, 2006, 90, 318-327.

[199] Morris, RL; Hollenbeck, PJ. J. *Cell. Sci.*, 1993, 104, 917-927.

[200] Wu, X; Hammer, JA. *Pigm. Cell. Res.*, 2000, 13, 241-247.

[201] Valetti, C; Wetzel, DM; Schrader, M; Hasbani, MJ; Gill, SR; Kreis, TE; Schroer, TA. *Mol. Biol. Cell.*, 1999, 10, 4107-4120.

[202] Waterman-Storer, CM; Karki, SB; Kuznetsov, SA; Tabb, JS; Weiss, DG; Langford, GM; Holzbaur, EL. Proc. *Natl. Acad. Sci. USA.*, 1997, 94, 12180-12185.

[203] Leopold, PL; Snyder, R; Bloom, GS; Brady, ST. *Cell. Motil. Cytoskel.*, 1990, 15, 210-219.

[204] Murray, JW; Bananis, E; Wolkoff, AW. Mol. *Biol. Cell.*, 2000, 11, 419-433.

[205] Wacker, I; Kaether, C; Kromer, A; Migala, A; Almers, W; Gerdes, HH. *J. Cell. Sci.*, 1997, 110, 1453-1463.

[206] Lopez de Heredia, M; Jansen, RP. *Curr. Opin. Cell. Biol.*, 2004, 16, 80-85.

[207] Smith, GA; Gross, SP; Enquist, LW. Proc. *Natl. Acad. Sci. USA.*, 2001, 98, 3466-3470.

[208] Suomalainen, M; Nakano, MY; Keller, S; Boucke, K; Stidwill, RP; Greber, UF. *J. Cell. Biol.*, 1999, 144, 657-672.

[209] McDonald, D; Vodicka, MA; Lucero, G; Svitkina, TM; Borisy, GG; Emerman, M; Hope, TJ. *J. Cell. Biol.*, 2002, 159, 441-452.

[210] Shah, JV; Flanagan, LA; Janmey, PA; Leterrier, JF. *Mol. Biol. Cell.*, 2000, 11, 3495-3508.

[211] Gross, SP. *Phys. Biol.*, 2004, 1, R1-R11.

[212] Welte, MA. *Curr. Biol.*, 2004, 14, R525-R537.

[213] Lipowsky, R; Chai, Y; Klumpp, S; Liepelt, S; Müller, MJ. *I. Physica A.*, 2006, 372, 34-51

[214] Hong, W; Shuo-Xing, D; Peng-Ye, W. *Chinese. Phys. Lett.*, 2008, 25, 321-324.

[215] Huxley, AF; Simmons, RM. *Nature.*, 1971, 233, 533-538.

[216] Dennis, JR; Howard, J; Vogle, V. *Nanotechnology*, 1999, 10, 232–236.

[217] Hess, H; Clemmens, J; Qin, D; Howard, J; Vogel, V. *Nano Lett.*, 2001, 1, 235-239.

[218] Hess, H; Clemmens, J; Matzke, C. M; Bachand, GD; Bunker, BC; Vogle, V. *Appl. Phys A-Mater.*, 2002, A75, 309-313.

[219] Clemmens, J; Hess, H; Howard, J; Vogle, V. *Langmuir.*, 2003, 19, 1738-1744.

[220] Suzuki, H; Yamada, A; Oiwa, K; Nakayama, H; Mashiko, S. *Biophys. J.*, 1997, 72, 1997-2001.

[221] Nicolau, DV; Suzuki, H; Mashiko, S; Taguchi, T; Yoshikawa, S. *Biophys. J.*, 1999, 77, 1126-1134.

[222] Wright, J; Pham, D; Mahanivong, C; Nicolau, DV; Kekic, M; Remedios, CGD. *Biomed. Microdevices.*, 2002, 4, 205-211.

[223] Hiratsuka, Y; Tada, T; Oiwa, K; Kanayama, T; Uyeda, TQP. *Biophys. J.*, 2001, 81, 1555-1561.

[224] Mahanivong, C; Wright, JP; Kekic, M; Pham, DK; Remedios, CGD; Nicolau, DV. Biomed. *Microdevices.*, 2002, 4, 111-116.

[225] Bunk, R; Klinth, J; Montelius, L; Nicholls, IA; Omling, P; Tagerud, S; Mansson, A. Biochem. *Biophys. Res. Commun.*, 2003, 301, 783-788.

[226] Moorjani, SG; Jia, L; Jackson, TN; Hancock, WO. *Nano Lett.*, 2003, 3, 633-637.

[227] Bohm, KJ; Stracke, R; Baum, M; Zieren, M; Unger, E. *FEBS. Lett.*, 2000, 466, 59-62.

[228] Riveline, D; Ott, A; Julicher, F; Winkelmann, DA; Cardoso, O; lacapere, JJ; Magnusdottir, S; Viovy, JL; Gorre-Talini, L; Prost, J. Eur. Biophys. J. *Biophys. Lett.*, 1998, 27, 403-408.

[229] Asokan, SB; Jawerth, L; Carroll, RL; Cheney, RE; Washburn, S; Superfine, RR; *Nano Lett.*, 2003, 3, 431-437.

[230] Stracke, R; Bohm, KJ; Wollweber, L; Tuszynski, JA; Unger, E. Biochem. Biophys. Res. *Commun.*, 2002, 293, 602-609.

[231] Felgner, H; Frank, R; Bierant, J; Mandelkow, EM; Mandelkow, E; Ludin, B; Matus, A; Schliwa, M. *J. Cell. Biol.*, 1997, 138, 1067-1075.

[232] Prots, I; Stracke, R; Unger, E; Bohm, K. *J. Cell. Biol. Int.*, 2003, 27, 251-253.

[233] Bohm, KJ; Stracke, R; Muhling, P; Unger, E. *Nanotechnology.*, 2001, 12, 238-244.

[234] Limberis, L; Stewart, RL. SPIE Int. Soc. *Opt. Eng.*, 1998, 3515, 66.

[235] Spudich, JA; Korn, SJ; Sheetz, MP. *Nature.*, 1985, 315, 584-586.

[236] Limberis, L; Magda, JJ; Stewart, RJ. *Nano Lett.*, 2001, 1, 277-280.

[237] Limberis, L; Stewart, RJ. *Nanotechnology.*, 2000, 11, 47-51.

[238] Langford, GM. *Curr. Opin. Cell. Biol.*, 1995, 7, 82-88.

[239] Lalli, G; Gschmeissner, S; Schiavo, G. *J. Cell. Sci.*, 2003. 116, 4639-4650.

[240] Morris, RL; Hollenbeck, PJ. *J. Cell. Biol.*, 1995, 131, 1315-1326.

[241] Lakadamyali, M; Rust, MJ; Babcock, HP; Zhuang, X. Proc. *Natl. Acad. Sci. USA.*, 2003, 100, 9280-9285.

[242] Apodaca, G. *Traffic.*, 2001, 2, 149-159.

[243] Goode, BL; Drubin, DG; Barnes, G. *Curr. Opin. Cell. Biol.*, 2000, 12, 63-71.

[244] Khataee, AR; Vatanpour, V; Amani Ghadim, AR. *J. Hazard. Mater.*, 2009, 161, 1225-1233

[245] Mavroidis, C; Dubey, A; Yarmush, ML. *Ann. Biomed. Eng.*, 2004, 6, 363-395.

[246] Mavroidis, C; Dubey, A. *Nat. Mater.*, 2003, 2, 573-574.

[247] Balzani, R; Credi, V; Gandolfi, A; Venturi, MT. Acc. *Chem. Res.*, 2001, 34: 445-455.

[248] Hamdi, M; Ferreira, A; Sharma, G; Mavroidis, C. *Microelectr. J.*, 2008, 39, 190-201.

[249] Hamdi, M; Sharma, G; Ferreira, A; Mavroidis, D. IEEE International Conference on *Robotics and Biomimetics.*, 2005, 105-110.

[250] Knoblauch, M; Peters, WS. Cell. Mol. *Life Sci.*, 2004, 61, 2497-2509.

[251] Hess, H; Vogel, V. *J. Biotechnol.*, 2001, 82, 67-85.

[252] Clemmens, J; Hess, H; Doot, R; Matzke, CM; Bachand, GD; Vogel, V. *Lab Chip.*, 2004, 4, 83-86.

[253] Astier, Y; Bayley, H; Howorka, S. Curr. *Opin. Chem. Biol.*, 2005, 9, 576-584.

[254] Nicolau, DV; Nicolau Jr., DV; Solana, G; Hanson, KL; Filipponi, L; Wang, L; Lee, A. P. *Microelectron. Eng.*, 2006, 83, 1582-1588.

[255] Spudich, JA. *Nature.*, 1994, 372, 515-518.

[256] Hess, H; Bachand, GD; Vogel, V. *Chemistry.*, 2004, 10, 2110-2116.

[257] Lavan, DA; Mcguire, T; Langer, R. *Nat. Biotechnol.*, 2003, 21, 1184-1191.

[258] Schmidt, JJ; Montemagno, CD. *Annu. Rev. Mater. Res.*, 2004, 34, 315-337.

[259] Suh, J; Dawson, M; Hanes, J. Adv. *Drug. Deliver. Rev.*, 2005, 57, 63-78.

[260] Cohen, RN; Rashkin, MJ; Wen, X; Szoka Jr, FC. *Drug. Discov. Today.*, 2005, 2, 111-118.

[261] Gunawardena, S; Goldstein, LSB. *J. Neurobiol*, 58, 258-271, 2004.

[262] Hurd, DD; Saxton, WM. *Genetics*, 1996, 144, 1075-1085.

[263] Goldstein, LSB. Proc. Nat. *Acad. Sci.* USA., 2001, 98, 6999-7003.

[264] Lipowsky, R; Klumpp, S; Nieuwenhuizen, TM. *Phys. Rev. Lett.*, 2001, 87, 108101.

[265] Klumpp, S; Lipowsky, R. *J. Stat. Phys.*, 2003, 113, 233-268.

[266] Klumpp, S; Nieuwenhuizen, TM; Lipowsky, R. *Biophys. J.*, 2005, 88, 3118-3132.

[267] Bunk, R; Klinth, J; Rosengren, J; Nicholls, I; Tagerud, S; Omling, P; Mansson, A; Montelius, L. *Microelectron. Eng.*, 2003, 67-68, 899-904.

[268] Hess, H; Bachand, G. D. *Nanotoday.*, 2005, 8, 22-29.

[269] Sharma, G; Badescu, M; Dubey, A; Mavroidis, C; Tomassone, SM; Yarmush, ML. *J. Mech. Design.*, 2005, 127, 718-727.

[270] Daneshvar, N; Ayazloo, M; Khataee, AR; Pourhassan, M. *Bioresource Technol.*, 2007, 98, 1176-1182.

[271] Khataee, AR; Daneshvar, N; Rasoulifard, MH. 1st Seminar on Nanotechnology *Applications and concept.*, 2004, 1.

[272] Brunner, C; Ernst, KH; Hess, H; Vogel, V. *Nanotechnology.*, 2004, 15, S540-S548.

[273] Marden, JH; Allen, LR. Proc. *Natl. Acad. Sci. USA.*, 2002, 99, 4161-4166.

INDEX

A

AAA, 9, 22
accuracy, 17
actin, 5, 6, 7, 8, 11, 12, 13, 14, 25, 27, 28, 30, 32, 40, 43
active transport, ix, 13, 36, 43
actuation, 40
adenosine, 2
administration, 40
ADP, 2, 16, 18, 19, 20, 21, 26, 28
agents, 38, 43
alternative, 37, 39
application, x, 25, 36, 38, 40
ATP, ix, 2, 8, 9, 10, 11, 13, 15, 16, 18, 19, 21, 22, 23, 26, 27, 28, 32, 33, 36, 37
ATPase, 9, 15, 21, 22, 29
attachment, 16
availability, 41
averaging, 29
axon, 6, 7
axonal, 32, 39

axons, 6

B

bacteria, 1
bacterial, ix
barriers, 39
BART, 39, 41
behavior, 5, 24, 26, 40
bending, 23
benefits, 1
binding, 5, 7, 9, 10, 11, 13, 15, 16, 19, 22, 23, 24, 26, 27, 28, 29, 32, 33, 39, 41
bindings, 16
bioengineering, 40
biological macromolecules, 36
biological processes, 41
biomimetic, 39
biomolecular, 39, 41
biomolecules, 40
bipolar, 27
bottom-up, ix, 36